Contributions to Statistics

V. Fedorov/W.G. Müller/I. N. Vuchkov (Eds.) Model-Oriented Data Analysis,
XII/248 pages, 1992

J. Antoch (Ed.) Computational Aspects of Model Choice,
VII/285 pages, 1993

W. G. Müller/H. P. Wynn/A. A. Zhigljavsky (Eds.)
Model-Oriented Data Analysis,
XIII/287 pages, 1993

P. Mandl/M. Hušková (Eds.)
Asymptotic Statistics
X/474 pages, 1994

P. Dirschedl/R. Ostermann (Eds.)
Computational Statistics
VII/553 pages, 1994

C. P. Kitsos/W. G. Müller (Eds.)
MODA4 – Advances in Model-Oriented Data Analysis,
XIV/297 pages, 1995

H. Schmidli
Reduced Rank Regression,
X/179 pages, 1995

W. Härdle/M. G. Schimek (Eds.)
Statistical Theory and Computational Aspects of Smoothing,
VIII/265 pages, 1996

Sigbert Klinke

Data Structures for Computational Statistics

With 108 Figures
and 43 Tables

Springer-Verlag Berlin Heidelberg GmbH

Series Editors
Werner A. Müller
Peter Schuster

Author
Dr. Sigbert Klinke
Humboldt-University of Berlin
Department of Economics
Institute of Statistics and Econometrics
Spandauer Str. 1
D-10178 Berlin, Germany

ISBN 978-3-7908-0982-4

Cataloging-in-Publication Data applied for
Die Deutsche Bibliothek – CIP-Einheitsaufnahme
Klinke, Sigbert: Data structures for computational statistics: with 43 tables / Sigbert Klinke. – Heidelberg: Physica-Verl., 1997
(Contributions to statistics)
ISBN 978-3-7908-0982-4 ISBN 978-3-642-59242-3 (eBook)
DOI 10.1007/978-3-642-59242-3

Originally published by Physica-Verlag Heidelberg in 1997

Softcover Design: Erich Kirchner, Heidelberg

SPIN 10558916 88/2202-5 4 3 2 1 0 – Printed on acid-free paper

Preface

Since the beginning of the seventies computer hardware is available to use programmable computers for various tasks. During the nineties the hardware has developed from the big main frames to personal workstations. Nowadays it is not only the hardware which is much more powerful, but workstations can do much more work than a main frame, compared to the seventies. In parallel we find a specialization in the software. Languages like COBOL for business-orientated programming or Fortran for scientific computing only marked the beginning. The introduction of personal computers in the eighties gave new impulses for even further development, already at the beginning of the seventies some special languages like SAS or SPSS were available for statisticians.

Now that personal computers have become very popular the number of programs start to explode. Today we will find a wide variety of programs for almost any statistical purpose (Koch & Haag 1995).

The past twenty years of software development have brought along a great improvement of statistical software as well. It is quite obvious that statisticians have very specific requirements for their software. There are two developments in the recent years which I regard as very important. They are represented by two programs:

- the idea of object orientation which is carried over from computer science and realized in S-Plus
- the idea of linking (objects) is present since the first interactive statistical program (PRIM-9). In programs like DataDesk, X-Lisp-Stat or Voyager this idea has reached its most advanced form. Interactivity has become an important tool in software (e.g. in teachware like CIT) and statistics.

The aim of this thesis is to discuss and develop data structures which are necessary for an interface of statistics and computing. Naturally the final aim will be to build powerful tools so that statisticians are able to work efficiently, meaning a minimum use of computing time.

Before the reader will read the details, I will use the opportunity to express my gratefulness to all the people who helped me and joined my way. At the first place is, Prof. Dr. W. Härdle. Since 1988 when I started to work as a student for him he guided me to the topic of my thesis. The development of **XploRe** 2.0, where I had only a small participation, and **XploRe** 3.0 to 3.2 gave me a lot of insights in the problems of statistical computing. With his help I got a grant from the European Community, which brought me to Louvain-la-Neuve and to my second Ph.D.-advisor, Prof. Dr. L. Simar.

A lot of people from CORE have contributed to the accomplishment of my work. I would like to mention Heraclis Polemarchakis, Luc Bauwens and Sheila Verkaeren. I am very thankful to the staff of the "Institut de Statistique" for their support and help, especially Leopold Simar, Alois Kneip, Irene Gijbels and Alexander Tsybakov. The atmosphere of Louvain-la-Neuve was very inspiring for my work. I have to mention the conference about "Statistical Computing" hold in Reisensburg because it gave me an insight in a lot of practical problems which have enriched my thesis.

I have also to thank a lot of friends and colleagues for their help and company: Isabel Proenca, Margret Braun, Berwin Turlach, Sabine Dick, Janet Grassmann, Marco and Maria Bianchi, Dianne Cook, Horst and Irene Bertschek-Entorf, Dirk and Kristine Tasche, Alain Desdoigt, Cinzia Rovesti, Christian Weiner, Christian Ritter, Jörg Polzehl, Swetlana Schmelzer, Michael Neumann, Stefan Sperlich, Hans-Joachim Mucha, Thomas Kötter, Christian Hafner, Peter Connard, Juan Rodriguez, Marlene Müller and of course my family.

I am very grateful for the financial support of the Deutsche Forschungsgemeinschaft (DFG) through the SFB 373 "Quantifikation und Simulation ökonomischer Prozesse" at the Humboldt University of Berlin which makes the publication of my thesis possible.

Contents

1
Introduction

Summary

This chapter first explains what data structures are and why they are important for statistical software. Then we take a look at why we need interactive environments for our work and what the appropriate tools should be. We do not discuss the requirements for the graphical user interface (GUI) in detail. The last section will present the actual state of soft- and hardware and which future developments we expect.

1.1 Motivation

What are data structures ?

The term "Data Structures" describes the way how data and their relationships are handled by statistical software. Data does not only mean data in the common form like matrices, arrays etc, but also graphical data (displays, windows, dataparts) and the links between all these data. This also influences the appearance of a programming language and we have to analyze this to some extent too.

Why examining data structures ?

In statistical software we have to distinguish between two types of programs: programs which can be extended and programs which only allow what the programmer had intended. In order to extend the functionality of the programs of the first class we would need a programming language which can not be recognized by the user (e.g. visual programming languages). This is important for statistical research, if we want to develop new computing methods for statistical problems.

We have a lot of successful statistical software available, like `SAS`, `BDMP`, `SPSS`, `GAUSS`, `S-Plus` and many more. Mostly the data structure is developed ad hoc, and the developers have to make big efforts to integrate new developments from statistics and computer science. Examples are the inclusion of the Trellis display or the linking facilities in `S-Plus` or the interactive graphics in `SAS`.

Therefore it seems necessary to decompose the tools of a statistical program (graphics, methods, interface) and to see which needs statisticians have and to develop and implement structures which in some sense will be general for all statistical programs.

Nevertheless some decisions are depending on the power of the underlying hardware. These will be revised as soon as the power of the hardware increases.

The programs of the second class can hide their structures. An analysis of these programs will be very difficult. We can only try to analyze the data structure by their abilities and their behaviour.

What is my contribution?

We first examine the needs in graphics, linking and data handling in extendable statistical software. The next step is to develop data structures that allow us to satisfy the needs as well as possible. Finally we describe our implementation of the data structures. There was a discrepancy between our ideas and the implementation in `XploRe` 3.2, partly due to the fact that this implementation exists longer than my thesis, but we also had some technical limitations from the side of the hard- and software. For example, in the beginning we had a 640 KB-limit of the main memory and we did not use Windows 3.1 in `XploRe` 3.2. In `XploRe` 4.0, under UNIX, we will implement our ideas in a better way, but we are still at the beginning of the development.

A extendable statistical software is composed of three components:

- the graphical user interface (GUI)
  In the first chapter we discuss the GUI shortly regarding why we need interactive programmable environments.
- the statistical graphic
  The graphics have to fulfill certain goals: there are statistical graphical methods and we need to represent the results of our analysises. So in chapter 2 we examine statistical graphics, in chapter 3 and 4 complete statistical methods (exploratory projection pursuit, cluster analysis) will be discussed.
- the statistical methods
  The statistical methods are often difficult to separate from the graphics (e.g. grand tour, exploratory projection pursuit). However we can decompose graphical objects into a mathematical computation and into a visualization step. We show this in the beginning of chapter 5. Another aspect of statistical methods is the deepness of the programming language. The deepness for regression methods is discussed in detail in the last section of chapter 4.

Part of the programming language is also the handling of data objects. In chapter 5 we give two examples (local polynomial regression, exploratory projection pursuit) why matrices are not sufficient for data handling. The use of arrays has consequences for the commands and the operators in a programming language. The need for hierarchical objects to store different objects and metadata also has an impact on commands and operators.

The question of linking data objects (data matrices, graphical objects etc.) is also part of chapter 5. The last chapter describes the implementation in the software `XploRe`. In `XploRe` 3.2, a program which runs under DOS, we have implemented the data structures of graphics and linking. In `XploRe` 4.0, which currently runs under UNIX and Motif, we have implemented arrays.

Where are the difficulties ?

The implementation phase of `XploRe` 3.2 lasted of course more than two years. The main problem at the beginning was that I did not have any idea which needs a statistician has. Nevertheless the decision about the data structures had to be made in an early phase of the implementation. Each missdecision I made had to be corrected later with an enormous amount of work. Some examples are:

- the programming language
  When I developed the programming language my aim was to build a language which simplifies matrix manipulations, but I did not want to develop a whole language with loops, selections etc. So I chose to build an interpreter which makes the `XploRe` language slow. Especially loops which interpret each line again and again instead of interpreting it once by using a compact code are very slow.

- the basic datatype
  For a matrix of numbers I chose the 4-byte float numbers as a basic datatype. Since in the beginning we had a technical limitation under DOS with max. 640 KB RAM, we wanted to store float numbers as short as possible. Since the compiled program already needs 400 KB memory we were only able to handle small datasets. Later I figured out that for some computations the precision was not high enough, so I changed to 8-byte float numbers. It took me some months to get the program to run properly afterwards.

- linking and brushing
  The implementation of linking and brushing in `XploRe` allows only transient brushing. This is due to the data structure I chose. After recognizing this I decided it was not worthwhile implementing a structure which allows nontransient brushing in `XploRe` 3.2. With an additional structure this would be possible, and we will correct this decision in `XploRe` 4.0.

The data structure I present in chapter 5 appeared only after I had thought about the needs. In fact it was a process of trying and learning. When Isabel Proenca implemented the teachware macros I saw that we needed programmable menus. So I implemented the command **MENU** which works in a window. One problem is that the cursor was supposed to change from an arrow to a bar. But after printing a display with **<Ctrl-p>** the cursor again appeared as an arrow. Only after the next movement it would appear as bar again. Another problem appeared when printing the **<F3>**-box together with boxplots. The standard behaviour was that the box disappeared from the screen and reappeared after printing, but did not appear in the printout. It took me nearly a week to change this behaviour.

Nevertheless I believe that I could build an efficient system. The development of most of the tools took me only one day or two. Of course the fine tuning like linking and everything appearing in the right time and the right place often took much more time. The wavelet regression macro is an example for this: the development was done in one day, but for the final form I needed more than a week. Additionally to the inclusion of the threshold method suggested by Michael Neumann I had to modify the commands handling the wavelet computations.

The analysis of data structures in other programs is very difficult. Since most them are commercial products I have no access to the source codes. Only from the way how the user sees these programs I can try to guess which data structures are used. Dynamic graphics and linking seems to be a real problem in **S-Plus** (e.g. there is practically no possibility for printing the scatterplot matrix). New tools like the Trellis display or linking require a quite extended reprogramming of these languages. So I only give a short overview of the facilities of the different programs.

Another problem was that I needed an overview about a lot of different statistical techniques, and I needed knowledge about the implementation of these techniques rather than the statistical and theoretical properties.

Some interesting problem, e.g. the treatment of missings or the error handling, could not be handled in a proper way because of a lack of time.

1.2 The Need of Interactive Environments

1.2.1 Why Interactivity ?

As soon as interactive graphical terminals were available statisticians start to use them. In 1974, Fisherkeller, Friedman & Tukey (1988) developed a program called **PRIM-9**, which allowed analyzing a dataset of up to nine dimensions interactively. They implemented the first form of linking, e.g. they allowed to mask out datapoints in one dimension such that all datapoints above or below a certain value would not be drawn. In a scatterplot which shows two different variables of the dataset the according datapoints would also not be drawn. They showed (with an artificial dataset) that this can lead to new insights about the dataset.

The computational equipment which was used in the seventies to run PRIM-9 was expensive. Nowadays computer power has improved and a lot of programs offer the facilities of PRIM-9. Nevertheless the idea of interactive environments needs time to become a standard tool in statistics.

In batch programming as it was common during the sixties and seventies, a statistical analysis needed a lot of time. There were two possibilities to work: step by step, which consumes a lot of time, or to write big programs which compute everything. The programming environment SAS, a program available already for a long time, computes a lot of superfluous informations although we may be interested just in the regression coefficients. As an example we show the regression of the variable FA (area of a flat in square meter) against FP (the price per flat in thousand DM) of the Berlin flat data; for a description of the dataset see section A. Figure 1.1 shows the SAS-output for a linear regression.

Figure 1.2 is an indicator of the analysis on the run, which shows that the linear regression ($y = ax + b$) is not appropriated for this dataset. Especially at the ends of the dataset we would expect another behaviour. For the left end we can see this from the coefficients ($a \sim 5, b \sim -67$), which tells us we would even get money for a flat with less than 13 m^2 if we wanted to buy it. A typical behaviour is to use $\log(FP)$ instead of FP. So we can use, as in this example, the interactive graphics to control the analysis. If we are not satisfied with the analysis we have to interfere. Here we will have to choose a nonlinear or nonparametric model. Users also like to control an analysis since they do not trust computers too much. An example might be that different statistical programs will give slightly different answers although they perform the same task (rounding errors, different algorithms).

Interactivity offers to cover "uncertainty" or nonmathematics. Uncertainty means that we do not know which (smoothing) parameter to choose for a task; see for example the smoothing parameter selection problem in exploratory projection pursuit. Often we can simply ask a user for the smoothing parameter, because he has a feeling for the right values for the parameter or can figure it (interactively) out.

Sometimes it is difficult to represent a visual idea as a mathematical formula. As example serves the selection of a clustering algorithm (hierarchical methods: the choice of the distance, the choice of the merging method). We have no general definition of a cluster, and as a consequence we have lot of

```
     The SAS System        17:09 Monday, July 31, 1995   2

Model: MODEL1
Dependent Variable: FP

     Analysis of Variance

                          Sum of          Mean
  Source          DF     Squares         Square    F Value  Prob>F

  Model            1 61329361.148 61329361.148   4686.772  0.0001
  Error         1365 17861883.086 13085.628635
  C Total       1366 79191244.234

       Root MSE      114.39243     R-square         0.7744
       Dep Mean      357.73167     Adj R-sq         0.7743
       C.V.           31.97716

     Parameter Estimates

                    Parameter       Standard     T for H0:
Variable  DF         Estimate          Error   Parameter=0    Prob > |T|

INTERCEP   1      -67.531740      6.93971316        -9.731        0.0001
FA         1        5.429284      0.07930592        68.460        0.0001
```

FIGURE 1.1. Output of the linear regression of the variables FA and FP of the Berlin flat data.

different possibilities to find clusters in the data. Interactivity allows us to give our expectations into the process of clustering.

Another important advantage of interactivity is that we can "model" sequentially:

- In Figure 1.2 we made a linear regression. In fact we could try a lot of different regression models. One possibility would be to use nonparametric models; our model might not satisfy the user. Figure 1.4 shows a robust locally weighted regression. This method tells us something different about the dataset.
- If we compute the correlation coefficient r_{xy} between the two variables FA and FP we see in Figure 1.1 that it is ~ 0.77. Often programs make a test immediately afterwards if the coefficient is unequal zero. With such a large value of the correlation coefficient ($n = 1366$), no test

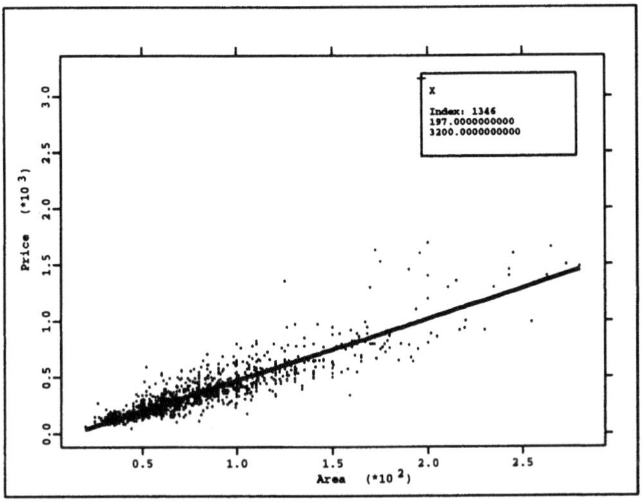

FIGURE 1.2. Output of the linear regression of the variable FA (area of flat) and FP (price of flat) of the Berlin flat data (only offers Oct. 1994).

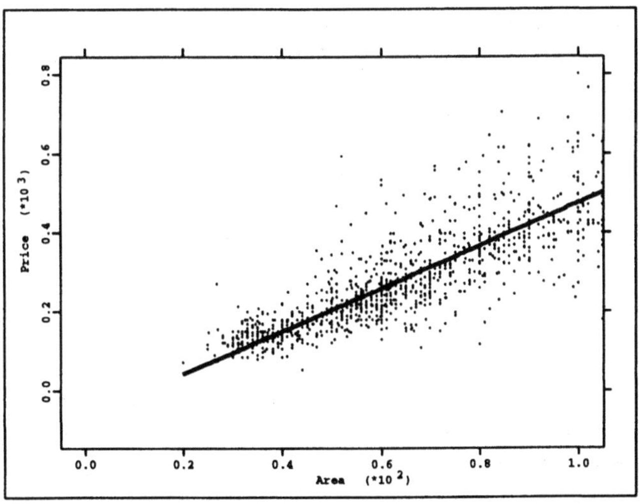

FIGURE 1.3. Same picture as Figure 1.2, but focussed in the left lower corner.

will accept the hypothesis $r_{xy} = 0$. This test only makes sense if the coefficient is near zero.

Interactivity also allows parallel modeling. For example we can make a linear and a nonlinear regression on our data, then we can analyze the residuals, the influence of observations on the fit etc parallel in both methods, e.g. via linking.

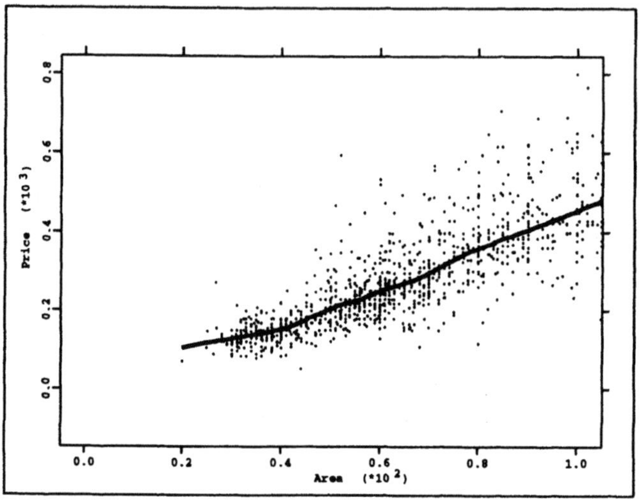

FIGURE 1.4. Robust locally weighted regression of the variables FP (price in thousand DM) against FA (area of a flat in square meters). This estimate coincides better with the feeling that the price should become constant or zero if the area decrease to zero.

1.2.2 The Tools of Interactivity

General

In the last five years the basic operating systems have changed. With the graphical capabilities available now, the interface of computers have changed from text based systems (DOS, C-Shell, Korn-Shell, VMS, TSO etc) to graphical interfaces. A lot of computers (MacIntosh, Atari) were developed which have only graphical interfaces, for other computers graphical interfaces were developed which sit on top or replace a text based interface (Windows 3.1, X-Windows, OS/2 2.11). Nowadays even operating systems are available independent of the underlying hardware (Windows NT, NextStep) and we see a development to unify even the hardware (PowerPC).

The advantage of the window based operating systems is their easiness of use: instead of having to remember a lot of different commands, we have just to click the (right) buttons, the menu items and the windows. This is a good choice for beginners. Nevertheless a graphical window system is not always the best choice. Take for example **XploRe 2.0**, a completely menu driven program. The data vectors are stored in workspaces. The addition of two vectors is a complicated task. We have first to click in the menu that we want to make an addition. Afterwards all possible workspaces are offered twice to figure out which vectors we want to add. It turns out that a lot of clicking and moving through menus is necessary, whereas a simple command like

```
y = x[,1]+x[,2]
```

is much easier just to be typed.

The lesson we can learn from this example is that both is necessary: a graphical and a text based interface.

Many of the statistical programs overemphasize one of the components; **DataDesk** and **XGobi** are completely based on graphical interfaces whereas **S-Plus**, **Minitab**, **GAUSS** and **SAS** emphasize too much the text based interface.

The underlying concept of a window system is the one of an office. A desk is the basis where several folders are spread. Windows represent the different task we are working on. As a real desk can be overboarded with piles of papers and books, the window system can have opened too many windows so that we loose the overview. Especially if the programs do not close their windows by themselves when inactive the number of windows increases fast. The size of the screens has become larger and larger: some years ago a 14 inch screen was standard, nowadays more and more computers start with 17 inch screens. This only delays the problem. Another solution which computer experts offer is a virtual desktop so that we only see a part of the desktop on the screen. Via mouseclicking or scrolling we can change from one part to another. Nevertheless a program has to use use intelligent mechanism to pop up and down windows.

Windows

We need to have different kinds of windows: Windows that allow us to handle graphics (2D- and 3D-graphics) and windows that handle text (editors, help systems). It is useful to incorporate text in a picture. In fact in all window systems we have only graphical windows, but some behave like text windows. The windows themselves have to perform some basic operations. Stuetzle (1987) described a system called **Plot Windows**, which uses a graphical user interface and proposed some operations:

- A window should to be moved easily to another location.

- A reshape of a window should be easy.
- Stuetzle wanted to have the possibility of moving windows under a pile of windows. Today we can iconify a window, that means to close a window and to place it somewhere as an icon so that we remember that we have a window. A mouseclick reopens the window.

Displays

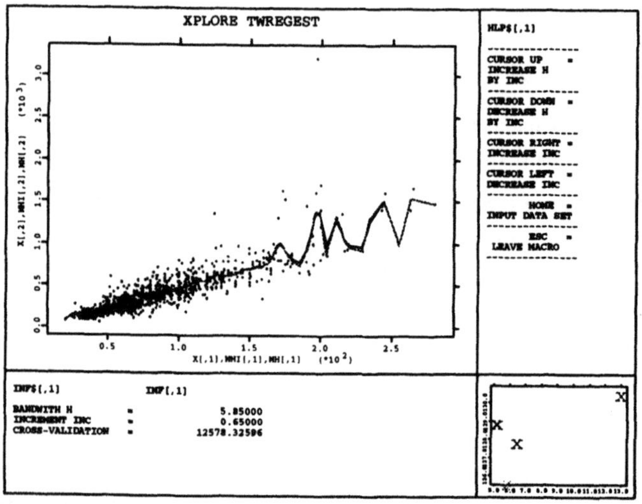

FIGURE 1.5. Windows with several informations form a display for a certain statistical task (here: relationship between the kernel regression and the selection of the smoothing parameter). A Nadaraya-Watson estimator is used to model the relationship between the variables FA (area of a flat) and FP (price of a flat).

From my point of view the concept of displays is important in statistics. In the windows of a display we can show several kinds of information that need to appear together. As an example see Figure 1.5, which shows a kernel regression on the Berlin flat dataset. The left upper window shows the data and the fitted line (gray), the right upper window gives us a small help text that tells us which keys can be used to manipulate the smoothing parameter (the bandwidth). The lower left window gives us information about the actual value of the smoothing parameter, the value increment or decrement of the smoothing parameter and the crossvalidation value, which can be used to find the smoothing parameter which fits best to the dataset. The last window, the lower right, shows the crossvalidation function for all smoothing parameters we have used. The aim of this macro is to teach students about kernel regression and crossvalidation, the whole set of the availabe macros is described in Proenca (1994).

All these windows belong to one statistical operation, and it hardly makes sense to show only a part of them. So a statistical programs can use this knowledge to perform automatic operations on a set of windows. A display will consist of a set of nonoverlapping windows, which belong to a certain statistical task. A display does not necessarily have to cover the whole screen as seen in Figure 1.5.

1.2.3 Menus, Toolbars and Shortcuts

Until now we have handled windows as static objects which do not change their contents. The example in Figure 1.5 needs some interaction from the user, mainly to change the bandwidth, the increase and the decrease of the bandwidth. In this example it is done through cursor keys, but in general menus are used for this. Normally menus appear at the upper border in the window and we can click on menu items to perform several tasks. Menus are diminishing the drawing area. On MacIntosh computers for example we have only one menu bar which is at the top of the window. The menu bar changes accordingly to the active window. One of the aims is to maximize the drawing area. A closer look to this solution shows that in the case of a big screen, this leads to long ways of the mouse, so it is reasonable to use pop up menus, which appear at the location of the mouse. This includes the possibility of fixing a menu on the screen if it will be used more often. In `XGobi` for example the "options" and the "print" menu are fixed menus and you have to dismiss them explicitly.

Menus are supposed to serve as a tool to simplify our work, especially if we have to do the same task again and again, e.g. a special analysis. We might want to extend the menu for our purposes. The underlying programming language has to have access to the menu structure in such a way that we can manipulate the menu. In `SPSS` for example we can make a scatterplot, and we have menuitems so that we can plot different regressions (linear, quadratic, cubic) in the plot. Nevertheless we miss the extensibility. In the teachware system of Proenca (1994) we are able to make a nonparametric smooth in a scatterplot. But the original macro does not include smoothing with wavelets, so we extended the macro. This means to extend the menu and to include the new method.

One drawback of the menus is that they require the choice of a language. All important statistical programs offer an "english" version, some are additionally extended to national languages since users prefer to communicate in their mother tongue. Although some efforts are made for the internationalization in programs, the problem still remains to translate a huge number of texts into many different languages. One approach how to solve the problem is the use of toolbars which means to use pictograms as menu items instead of words. Unfortunately we have no rules available how a pictogram for a certain task should look like. This leads to the problem that every software program which uses toolsbars will more or less use its own version of pictograms. Sometimes they are very different although the same operations are behind them. Another problem is that pictograms are usually small and it follows that they need a careful design, otherwise the user might not connect the pictogram to a certain operation.

Another drawback of the menus depend on the different type of users. Beginners and unexperienced users will very much like the choice by menus since they offer an easy access. But as mentioned above if we have to make long ways with the mouse to select some item the experienced user will get impatient. This results in the need of short-cuts, special key combinations, which will start the process behind a menu item as if we had clicked on the item. By frequent use of a program the users will learn the short cuts by heart and they can work very efficiently. Of course short-cuts are changing from program to program, e.g. `Ctrl-y` in Word, `ALT-k` in Brief, `Ctrl-K` in Emacs and so on to delete a line. This requires that the short-cuts are programmable too.

Interactive dialog

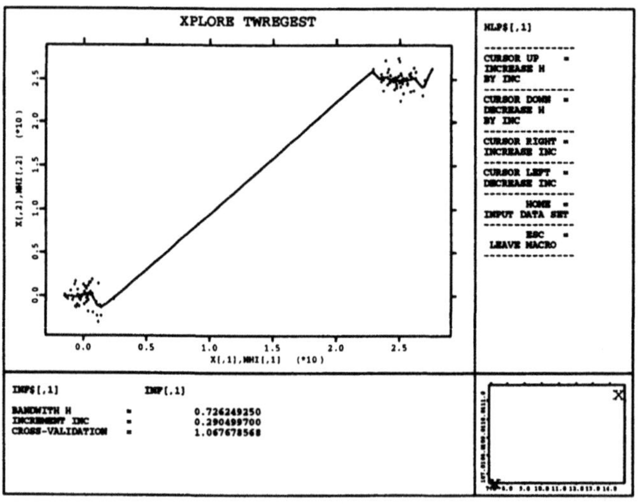

FIGURE 1.6. The automatic bandwidth choice for kernel regression leads to a completely missleading regression.

Communication between the user and the program via menus or toolbars is restricted to start a process. Sometimes it is required to give parameters to a routine. In the example of the kernel regression in Figure 1.6, a starting bandwidth has to be chosen. This is done in an automatically and the bandwidth is chosen as five times the binwidth. The binwidth is defined as

$$\text{binwidth} = \frac{\max_i X_i - \min_i X_i}{100}.$$

This choice is not always appropriate as Figure 1.6 shows. The problem arises that the bandwidth is far to small to cover the gap between both clusters of data. Of course if the bandwidth would be large enough it will oversmooth the data heavily. If the starting bandwidth could be chosen interactively, we would make a choice which would try to balance both effects. Dialog boxes which can be used to obtain a starting bandwidth appear all over in windows based systems, e.g. to read and write programs and data to the harddisk. In general, wherever parameter are required we can use dialogboxes.

Sometimes we will need special boxes with pictures, e.g. if we have to choose a colour. **S-Plus** offers an number which indicates an entry into a colour palette. The palette itself can be manipulated via a dialog box which does not include a immediate view of the colours. Here the user can choose colours by name or by RGB-triplets with hexadecimal numbers. The first method requires knowledge about the available colours, which can be received by calling **showrgb**, the second method requires some experience to connect the RGB-triplets to the desired colours. In the **S-Plus** manuals they give the advise that the user should create a piechart which contains all possible colours so that we get an immediate impression what happens if we change the colour palette.

To vary the bandwidth in **XploRe** the cursor keys are used. It would be better to use (log-linear) sliders as in **DataDesk** or **XGobi**. This will allow a fast and smooth change of the bandwidth. The last group of dialog tools are message boxes which give informations to the user like warnings, errors and so on.

Interactive programs in general require short response times. The response time should not be longer than a few seconds $(2 - 4)$. The exact time will often depend on the sample size, a user will not expect (maybe wish) that a regression with 30.000 cases will be as fast as a regression with 30 cases. A better acceptance of long response times is achieved by the use of a statusline indicating how much of a certain task is already done. The Windows 3.1 system changes normally the form of the mouse cursor to show that it is occupied, but this will be not very helpful if the system is occupied for a longer time.

1.2.4 Why Environments ?

The aim of an environment is to simplify the task we have to do as much as possible.

The first step of the analysis is to read the data into the program which can be a difficult task. An example is the read routine of SPSS where we have the possibility to read data from ASCII files. SPSS distinguishes between two formats, a "fixed" format and a "free" format.

In both formats it is expected that each row contains one observation. In the fixed format it is additionally expected that the variables always start in a fixed row and stop in another row. If we see datafiles today we have mostly one or more spaces as delimiter between the variables in a row. But even if the data are in such a formatted state we have to give SPSS the rownumbers which means we simply have to count the lines of the datafile.

One may think that the free format will be more helpful if the data are not in fixed format. Unfortunately the version of SPSS which I had available uses a comma for decimal numbers instead of a point, so we had to leave the menu driven environment and to modify the program. We had to add the command

```
SET DECIMAL=DOT.
```

and to execute the program. Unfortunately for me this special option is not documented in the online help so I needed the help from our computer center to read a simple ASCII datafile. In contrast to that we have routines like in S-Plus which by a "simple" command allow to read a dataset:

```
x <- matrix(scan("bank2.dat"),byrow=T,ncol=6)
```

To read data is a task that we have to do again and again. In general there will be a lot of tasks we have to repeat during statistical analysis.

We are interested to make our analysis as fast as possible. If we have found our way to make some kind of standard analysis, we would like to fix this way so that it can be repeated easily.

We need libraries which contain all the tools we need. It should be easy to make the tools we need from these libraries. The implementation of new statistical methods requires already well known statistical techniques which can be composed from these libraries.

Again we need a programming language that allows us to compose our tools. A statistical software system should offer tools which are broad enough to do a specific task well, but it should not cover too much.

If we have a good environment we can concentrate on the statistical analysis instead on reading the data.

1.2.5 The Tools of Environments

Editors

An important tool in a programmable statistical software is the editor. It will be the main tool to write the programs, to view and to manipulate the data.

It has to be easy and comfortable to use. Some editors miss important features like a blockwise copy. The main problem with an editor is that we have to know which key combinations will execute a specific task. The standards are very different. Modern editors allow a redefinition of the key combination and already offer standard sets of key combinations (e.g. Word compatible, Emacs compatible etc).

Especially an editor has to show data in an appropriate way. If we want to display a data matrix it will be a good idea to use a spreadsheet as editor. This kind of editor is widely used in statistical software.

For a big number of cases or variables we need to regroup the variables and cases. However, the use of spreadsheets as editors for large datasets will causes difficulties. These difficulties will increase if we use multidimensional arrays.

Help system

Broad and complete help systems are necessary for the user. It is very helpful if the help systems are available online. For example it would be difficult to have the SAS-manuals always at hand.

We need a clear description of the data and the methods. The statistical methods can be very complicated. Often a software package allows to make tests for a certain task. As long as we know the tests we can easily check the underlying assumptions. But if we do not know the tests and can not find them in standard literature we can not be sure if one of the underlying assumptions is not violated and that we interpret the test results wrongly.

But the help system should offer more than just simple explanations. Modern software offers the possibility of topic orientated helps which means if we want to make regression it will inform us what kind of regression methods we have available in the software package. Such kind of hypertext systems can be found in statistical software, e.g. in GAUSS. The hypertext systems are developing independently from the statistical software as the help system under Windows 3.1 or the HTML language for the World-Wide-Web (WWW) shows.

Of course we need some context-sensitive help which will give us an appropriate help depending on the actual context. For example if we are in an interactive graphic window we are interested to know how to manipulate the graphic, but we are not interested to get the command overview.

A good help system would not only offer specific informations to a topic but try to give some more general help. For example it would be worthwhile not only to get all possible routines for the regression but also some background information about the definition, the (small/finite sample) behaviour etc. The paper documentation of `SAS` is a good example.

Programming language and libraries

As pointed out earlier, we need a programming language and a menu driven environment. The menu driven environment allows us to do standard tasks in an easy way.

A programming language is the basic method for the manipulation of the data (structures). We can build up menu driven environments and statistical tools to simplify our work. This is important for the scientific research.

This also aims at the different user groups:

- Researcher
  who needs full access to all possible methods and language elements. They will need to develop new methods and new techniques.
- Consultants
  who need a variety of tools which allow them to make their analysis efficiently. Sometimes they will need to compose new tools from the existing ones.
- Students
  who mainly need good and easy user-interface with a context-sensitive help system. They will prefer a clicking and drop-and-drag environment.

For detailed overview about the programs being appropriate for each user group see section 5.4.1.

Since we have different needs we have to implement a programming language: it has to allow that the user can do everything on a very basic level, e.g. matrix manipulations. But we need a macro language that allows us to build tools efficiently. These tools have to allow for different user groups to satisfy their needs. *We need a multilevel programming language.* We need the concept of loadable libraries and programs (macro). Similar tools can be put together in libraries so that we only have to load libraries to have a set of tools available to solve a task.

When we talk about a programming language, we always mean a typed programming language as in `GAUSS` or `S-Plus`. Graphical programming languages are possible, but we do not believe that they are powerful enough to allow efficient programming for a researcher. How detailed a programming language should be can be seen in section 3.3, where we discuss for the case of the regression methods whether a specific regression needs to be a command or a macro.

We have some fundamental operations in a programming language:

1. Flow control
   We need some flow control elements like loops and selections:
   (a) Unconditioned loop
       `do ... endo`
   (b) Enumeration loop
       `for (i=0; i<n; i++)`
   (c) Preconditioned loop
       `while (i<n) ... endo`
   (d) Postconditioned loop
       `do .. while (i<n)`
   (e) Selection by condition
       `if (i<n) ... elseif (i<2n) ... endif`
   (f) Selection by number
       `switch (i) { case 0: ... default: ...}`
2. Operators
   Operators are mainly used for calculations. As we are used to write `x + y` we have to provide such operators for user-friendlyness as well, but we could use procedures like `sum(x,y)` instead. We have two classes of operators:
   (a) Unary
       Unary operators have only one argument, e.g. unary minus, faculty etc.
   (b) Binary
       Binary operators have two arguments, e.g. plus, minus, multiplication etc.
3. Procedures
   The most powerful operations are procedures in programming languages. They provide us with some output parameter, a procedure name and some input parameters.

From computer science we got new developments in the design of the programming languages, e.g. object orientation. In fact `S-Plus` tries to follow these ideas, but my impression is that not many people are really using this features in statistics. This might be due to the fact that we are used to the procedural languages.

Important is the possibility to load functions and libraries from the disk. It allows us to build a collection of procedures and to put similar procedures together. `Mathematica` and `XploRe` are example for this. Sometimes libraries consist of precompiled objects which can be executed faster.

1.3 Modern Computer Soft- and Hardware

1.3.1 Hardware

What do we already have?

The speed of processors is still increasing. Table 1.1 compares the speed of common processors from the beginning of the eighties until 1995. The comparison has to be used with extreme care! The tests (Dhrystone/SPAC (= SPACSyncCalcFloatingPoint)) are quite different and the comparison (Motorola/Intel) is based on approximate values. Also different computers have been used and the environment, hard- and software, will have had some influence as well. The column $O(n)$ represents the size of a dataset for a calculation which depends linear on the size, e.g. the mean; the column $O(n^2)$ represents the size of a dataset for a calculation which depends quadratic on the size, e.g. the direct Nadaraya-Watson estimator[1].

Both the improvement of the processor speed and the increase of memory and storage space provide the ability to compute big statistical models. Projection pursuit methods which mainly depend on the optimization, demonstrate this very well. Without the use of the power of workstations interactive programs like `XGobi` would be impossible. In the section about exploratory projection pursuit we mentioned that the treatment of discrete data has to be quite different from the treatment of continuous data. Today we have the power and the memory to handle this case, but we have to give up the interactivity.

Interactive graphics and animation used in the scatterplot matrix, interactive contouring or exploratory projection pursuit have proven their worth for statistics. With the use of graphical terminals we have left (for viewing) the area of black-and-white pictures and started to use colour. It can easily be demonstrated that the usual 16 colours of a VGA-card are not enough to get

[1]For the literature see Nachtmann, (1987a, 1987b), Earp & Rotermund (1987), Schnurer (1992), Meyer (1994) and Meyer & Siering (1994)

Processor	Motorola	Intel	Dhrystone	SPAC	$O(n)$	$O(n^2)$
6502 (C 64)			0.037		1.5	40
Motorola 68000 (8 Mhz)	1.6		0.87		38	195
Motorola 68010 (10 Mhz)	1.8				79	280
Motorola 68020 (12,5 Mhz)	3.7				161	400
Motorola 68020 (16.7 Mhz)	4.7		4.7		205	450
Motorola 68020 (20 Mhz)	5.5				240	490
Motorola 68020 (30 Mhz)	7.0				305	550
Motorola 68030 (20 Mhz)	10.0				436	660
VAX 8600			7.14		315	560
Cray-1A (80 Mhz)			13.9		612	780
Cray-XMP (105 Mhz)			17.8		766	875
Intel 8088 (4.77 Mhz)		100	0.341		15	120
Intel 8088 (8 Mhz)		168	0.59		25	160
Intel 8086 (8 Mhz)		335			49	220
Intel 80286 (6 Mhz)		252	0.91		40	200
Intel 80286 (10 Mhz)		419			62	250
Intel 80386 (16 Mhz)		1342	4.5		198	445
Intel 80486 (33 Mhz)			22.758		1000	1000
Intel 80486DX2 (50 Mhz)			45.56		2000	1414
Intel 80486DX2 (66 Mhz)			60.3	1700	2640	1625
Intel Pentium (66 Mhz)				2200	3420	1850
Intel Pentium (90 Mhz)				2500	3880	1970
PowerMac MPC601 (60 Mhz)				1900	2950	1720
PowerMac MPC601 (66 Mhz)				3100	4810	2190
PowerMac MPC601 (80 Mhz)				3500	5440	2330

TABLE 1.1. Numerical power of processors.

a smooth gradation from (dark) blue to (light) red. In image plots we might want to use a blue to show the areas where the minima occur and a red for the areas where the maxima occur. One attempt to show four dimensional data is to code the fourth variable as colour, but if we have only 16 colours we are restricted to discrete variables. A continuous variable will loose a lot of structure if we represent their values with 16 or less colours. Nowadays producers of graphic cards allow a choice of $256^3 = 16.7$ million different colours.

One drawback of interactive graphics, animations and colours is that we are not able to reproduce them on the paper in a appropriate way. Although nowadays colour printers are available it is almost impossible to get publications with colour pictures in journals. This might change with new journals like the "Journal of Computational and Graphical Statistics". Nevertheless interactive graphics or animations with a big spectrum of colours on the screen can not be reproduced by paper journals. Electronical journals like InterStat in the WWW or a distribution on a CD-ROM might change this.

Although the entertainment industry has greatly improved the quality of soundreproduction (compare a PC-loudspeaker to a modern soundcard) it is rarely used in statistics. The main drawback here is that the eye plays the most important role in human perception. Wilson (1982a, 1982b) has used sound for exploratory data analysis. Since human soundperception is logarithmic, e.g. doubling the volume means an increase by factor 10, it would allow to explore data of big ranges, e.g. from 1 to 10^7 (e.g. census of city population in the United States).

Nonparametric estimation are often founded on asymptotic theory, see e.g. the problem of bandwidth choice in kernel estimation or multivariate estimation. So we need many observations to get an estimation we would trust. New memory media like optical disks and CD-ROMs will allow the storage of huge datasets. Nevertheless we will have some areas of research, especially in economics, where the number of observations will be small (e.g. in deriving results in nationwide economics).

What can we expect from the future ?

Surely we will see a further increase of processor speed, which will render more complicated and computer intensive statistical modeling possible. The graphical techniques and the graphical representation of the results will enlarge much more. The graphic systems will become faster (grand tour, animation) and more powerful (more colours).

In computer industry people work hard to make it possible to use the natural language for input and output. For modern soundcards it is not very difficult to produce output in natural language, still the input has big problems recognizing human language. The solution will need expert systems which can handle statistical problems in a proper way. Expert systems are available only for very limited statistical problems (e.g. `GLIMPSE` by Nelder) and we do not know any commercial package offering even a limited expert system.

Nevertheless knowledge based expert systems will be able to support the statistical analysis we have to do. They will give us a partner for our thoughts who speaks the same language and reduce our work to the barely statistical problem.

Many existing statistical algorithms can easily be parallelized (see e.g. nonparametric estimation etc.). As a consequence there will be a increasing need for multi-processor machines or a network-wide distribution of tasks. Especially the distribution of statistical tasks over a lot of machines in a network will increase the computational speed. One approach is done via the software **MMM**. The aim of the **MMM**-software, which is based on World Wide Web, is to collect informations about (statistical) methods. These methods can be implemented in different hardware platforms and programs. The program will convert the data from one format into another format. Moreover it will know where it can find special methods for the analysis. If these methods are freely available it can copy the methods and execute them on the user machine or alternatively execute them on the machine where the method can be found. For details see Oliver, Müller & Weigend (1995) or Krishnan, Müller & Schmidt (1995).

Three dimensional devices for input (camera and software, tomographs) and output (laser) will give us the chance to see data in a more realistic and natural view. Although spinning is a good possibility to represent datapoints it still makes trouble to rotate thousands or tenthousands of datapoints. Only special software systems are capable to rotate shells which come from the density estimation of three dimensional data (in real time).

We hope that these developments will result in more user-friendly computer systems and new statistical techniques.

1.3.2 Software

Advances in software technology have always had their impact in computational statistics. We have to distinguish between two groups of people: one which is using statistical software and another which is creating it. The big impact in statistical computation for the first group mainly comes from increasing hardware facilities. Only the change from text based environments to graphical user interfaces (GUI) had been of some relevance. Together with GUIs we have multitasking possibilities and often some kind of hardware independence. `S-Plus` serves as an example: In the first version it was only possible to open one graphic window, then restart `S-Plus` and open another graphic window, but the communication between both graphic windows was difficult. `S-Plus` has also developed to a multi-platform program (UNIX, Windows), so we do not need to care about the platform we have. This has some disadvantages: the object format for compiled programs is standardized under UNIX, but not under DOS/Windows. As a result we can program statistical tasks under UNIX in any programming language we want. Under

DOS/Windows only Watcom-compilers are supported which are rarely used.

The other important impact from software technology is the introduction of *object orientation*. Most statistical programs are still working with lots of single data matrices (see e.g. `XploRe 3.2`). The user has to know the relationship between these matrices by heart. Here `S-Plus` offers the possibility to have (hierarchical) objects.

Lets take as example the projection pursuit regression (PPR) in S-Plus. The minimal command would be

```
ppr <- ppreg (x,y,termno)
```

with **x** a multivariate variable and **y** the one dimensional response. **termno** gives the minimal number of terms which will be used. **ppr** itself would be an object which is a list of objects containing the subobjects

ypred the residuals of the fitted model

fl2 the squared residuals divided by all corrected sums of squares

alpha a matrix of projection directions used

beta a matrix of the weights for every observation y_i

z a matrix which contains $\alpha_i^T X$

zhat a matrix which contains the smoothed values $\sum_{i=1}^{termno} g_i(\alpha_i^T X)$

allalpha three dimensional array which contains the fitted alphas for every step

allbeta three dimensional array which contains the fitted betas for every step

esq contains the fraction of unexplained variance

esqrsp contains the fraction of unexplained variance for every observation.

The subobjects can be accessed via **ppr$z**. It is obvious that we do not have to handle a lot of single matrices, but the program will do it an easy way for us.

Another important fact is the use of classes in `S-Plus`. It allows us to define a bunch of data matrices and the methods to handle it in one object. With the **ppr** object we are not directly able to achieve the results graphically. We would need a `S-Plus` program:

```
ppr <- ppreg (x,y,termno)
matplot (ppr$z, ppr$zhat).
```

If we would define an object orientated class "projection pursuit regression", we could (re)define a function `print`, which exists for every `S-Plus` object, in such a way that we get a graphical result immediately.

These objects can be used to hide nonimportant informations, this is called *encapsulation*. The normal user will not be too interested how we did the plot and what else we have incorporated in the data. *inheritance* ensures that derived objects will have all abilities of the original object, so we can include such a projection pursuit regression class in a broader context like a teachware tool for regression which might be a class by itself.

Beside the statistical task to make our calculations we have to represent our results. Multimedia-documents are able to show different kinds of representation like text, tables, graphics and animation in parallel. Methods like *object linking and embedding* under Windows 3.1 will allow interactive working with documents. A change of a dataset in a `WORD`-document can call the statistical program to recompute the pictures for this new dataset. But we should keep in mind that we are here on the edge of the computational power. Methods as *object linking and embedding* are widely available today in operating systems, but often not used.

Virtual reality means that we replace the whole input we get by human perception by an artificial environment created by a computer. Today's research has been successful in replacing the information we get through our eyes. The hope is that we can represent (hierarchical) statistical objects by graphics. The virtual space of our objects should lead to an easier manipulation of the objects and an easier recognition of relations between the objects.

2
Exploratory Statistical Techniques

Summary

The first step is to analyze the statistical graphics in use, so we examine the descriptive statistics. Next is the boxplot, the Q-Q plot, the histogram, the regressogram, the barchart, the dotchart, the piechart, the 2D-scatterplot, the 2D-contour plot, the scatterplot matrix, the 3D-scatterplot, the 3D-contour plot, the Chernoff faces and the parallel coordinate plot. We use some of the tools of `XploRe` to examine the Berlin housing dataset which use the plots mentioned above. Finally we state that two kinds of windows are necessary, one which can draw points, lines and areas, another which can draw glyphs windows, i.e. the Chernoff faces, star diagrams and so on.

2.1 Descriptive Statistics

Some descriptive statistics.

The first step of an analysis of a dataset can be the computation of some descriptive statistics of the variables of the dataset. Such descriptive statistics of each variable are

Missings the number of missing values

Discreteness the number of different values the variable take

Mean the mean value

Variance the variance

Std Dev. the standard deviation

Minimum the minimum

Maximum the maximum

Range the difference between maximum and minimum

1. quartile the first quartile

Median a more robust central value than the mean

3. quartile the third quartile

Obviously we can compute more descriptive statistics, e.g. skewness and curtosis. Nevertheless this is a first approach to get an overview about the variables. The Tables 2.1 and 2.2 show the descriptive statistics for the whole Berlin flat data.

	T	FA	FL	FR	FB	FM	FP	FI	FE
Missings	0	0	0	0	0	0	0	0	0
Discreteness	22	1437	42	18	2	2	1703	3967	2
Mean		79.1		2.5			331	301.1	
Variance		1594.6		1.4			49438	39981.0	
Std Dev.		39.9		1.2			222	200.0	
Minimum	8907	20		1	0	0	38	38.0	0
Maximum	9410	510		11	1	1	3200	2718.5	1
Range		490		10	1	1	3162	2680.5	1
1. quartile	9104	52		2	0	0	180	164.9	0
Median	9210	69		2	1	0	269	244.3	0
3. quartile	9401	97		3	1	0	420	380.7	0

TABLE 2.1. Some descriptive statistics of the Berlin flat data for the variables T, FA, FL, FR, FB, FM, FP, FI and FE.

	DI	DF	DU	DR	DW	DS	DN	DB	TI
Missings	0	0	0	0	0	0	0	0	0
Discreteness	59	58	34	19	19	28	27	24	22
Mean									
Variance									
Std Dev.									
Minimum	8.5	0	0	6.0	0.3	2.6	25	3	5.4
Maximum	138.9	26.6	8.7	23.7	8.2	11.1	101	27.5	9.2
Range	130.5	26.6	8.7	17.7	8.0	8.5	76	24.5	3.7
1. quartile									
Median									
3. quartile									

TABLE 2.2. Some descriptive statistics of the Berlin flat data for the variables DI, DF, DU, DR, DW, DS, DN, DB and TI.

We used the macro `DESKSTAT` from `XploRe 3.2` to compute the Tables 2.1 and 2.2 while the empty entries were done by hand. A statistical program would need informations about the variables to decide which descriptive statistics could be computed. It makes no sense to compute some descriptive statistics for different types of variables (nominal, ordinal, metric, time). For example SPSS makes an approach to store information about a variable, but this is not enough to allow or permit different statistical operations. The program BDMP stores informations about the type of a variable and permits a statistical operation if it is appropriate. This may lead to some problems since sometimes the type of a variable is not completely clear.

Missings.

The treatment of missings is a problem in statistical software. We know the MANET program (Unwin 1995) for exploratory statistics which gives a clear indication about missings to the user. In most statistical software those observations are excluded from statistical operations which contain missings. Nevertheless we have general methods available which are able to handle missings (Schwab 1991). For special statistical operations, e.g. linear regression, we have also modified algorithms available which can treat missings appropriately. At least statistical software should indicate that we have missing values.

SPSS stores about a variable

- the name,
- the format,
- the coding values and
- the missing values.

SPSS notifies different kinds of missings (system-defined and several user-defined) reflecting the fact that missings might have different reasons. From my point of view it is not satisfying to produce an output line like in SPSS which tells us how many observations are included in a statistical operation, see Figure E.1.

Fortunately the Tables 2.1 and 2.2 show that the Berlin flat data have no missings.

Facts from the variables.

Additionally we can see from the tables that the dataset is very discrete. Although we have 14968 observations, we have only three variables which have more than 1000 different values: FA, FP and FI. We see moreover that we have only 42 different districts. The variables which describe the district can have $42 \times 22 = 924$ different values, but in fact between 20 and 60 are taken.

What can we learn from the single variable:

- It seems that the average size of a flat is between 69 (median) and 79 square meters. We will see that the modus is at 60 square meter. But we also have a flat of 510 square meters.
- The average room number is around 2. We expect that the big values in the size of a flat correspond to many rooms.
- Most of the flats have a balcony (more than 50%), less than 25% are classified as maisonette flats. Less than 25% of the flats are in the east part of Berlin. A deeper analysis will show that we have only 248 out of 14968 offers in the east part, so models about the prices in the east part does not seem reasonable to me.
- The descriptive statistics might be misleading, because some variables are depend on the time.

Table E.2 shows the absolute frequencies of offers on time and district.

2.2 Some Stratifications

Stratification after location and time. From Table E.2 and Table E.3 we see that the data for the east are sparse for all time periods. We can try to find a model for the west part and apply to the east part, but we have to be aware that the west and the east part are quite different.

Stratification after time. Table E.4 and Table E.5 show how many observations are falling into each time period. The values of the variables describe the discreteness of the variable, that means how many different values are taken by the variable. It seems strange that the variable DU (unemployment rate) at most time periods only takes two values. Remarkable is that all district variables for one time period can be regarded as an aggregation of the location variable FL in several directions. To model the price it might be enough to use the variable FL as nominal variable. The district variables can be used as an explanation why the model looks the way it does.

Stratification after recreation area. As expected Table E.6 and Table E.7 show that the size of the recreation area remains stable for most districts. We also see that we have a one-to-one relationship between the variables DR and DW (blue collar workers). Although the datapoints do not form a one-to-one relationship we are able to identify the value of the variable DW from the value of the variable DR for an observation and vice versa. The datapoints are jittered which means they are distributed around the true value in the center of a point cloud. The aim is to see how many datapoints are hidden behind one point in the plot.

Stratification after interest rate of the German Bundesbank. From Table E.8 and Table E.9 we can see that we have a one-to-one relationship between the variables TI (interest rate) and T (time of offer). Although the datapoints do not form a one-to-one relationship we are able to identify the value of the variable T from the value of the variable TI for an observation. The datapoints are jittered which means they are distributed around the true value in the center of a point cloud. The aim is to see how many datapoints are hidden behind one point in the plot.

2.3 Boxplots

Aim.

The boxplot is an useful tool to analyze univariate data. It gives the statistician informations about locality, spread and skewness of a dataset. In some sense it is the graphical analogue to the five number summary (minimum, 1. quartile, median, 3. quartile, maximum) of a dataset.

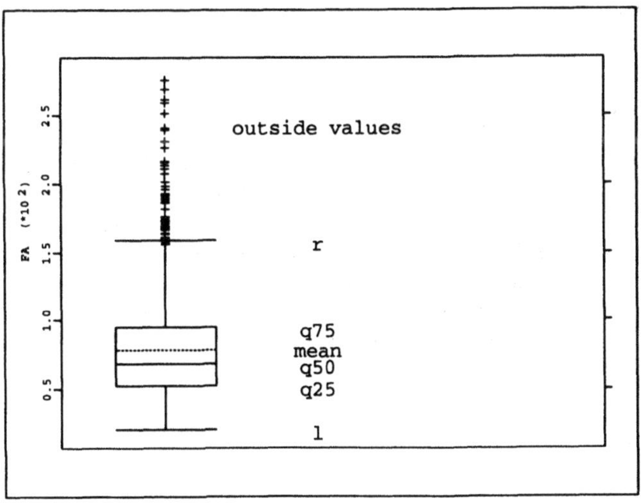

FIGURE 2.1 Construction of a boxplot (variable FA).

Construction.

Tukey (1970) gave the construction of a boxplot as in Figure 2.1. Here q_f is the f%-quantile of the dataset.

Datapoints left of l and right of r are called outside values which can be interpreted as outliers. It is obvious that not every outside value is an outlier, an example with nearly 50% outside values can easily be constructed.

McGill, Tukey & Larsen (1978) added features to the boxplots like using the width as a measure for the sample size and including notches as indicators for the (rough) significance of differences between medians. Further modifications of the visual appearance of the boxplots have been suggested by Tukey (1990).

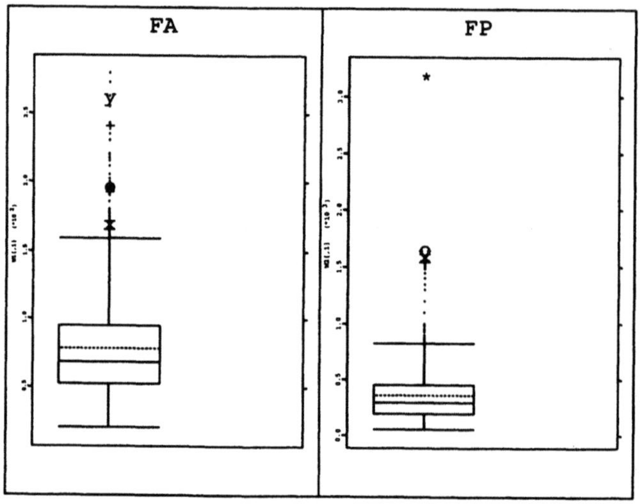

FIGURE 2.2. Linked boxplot of the variables FA and FP of the Berlin flat dataset (only offers from October 1994). The five upper outside values of FP are marked with different symbols. We can see that the flats with very high prices ($\sim 1.5 - 3$ million DM) correspond to the flats with a large area. The datapoint marked with a star shows a flat of nearly 200 m^2 and a price of approximately 3.1 million DM. Since the order of the marked values is quite different this gives us some statistical evidence that the price is not completely determined by the area.

Analyzing outside values.

Since boxplots are used to identify outside values, it is of interest to compare the variables of a dataset. The question rises if the outside values of one variable correspond to the outside values of another variable. To analyze this we need linked boxplots, as an example see Figure 2.2.

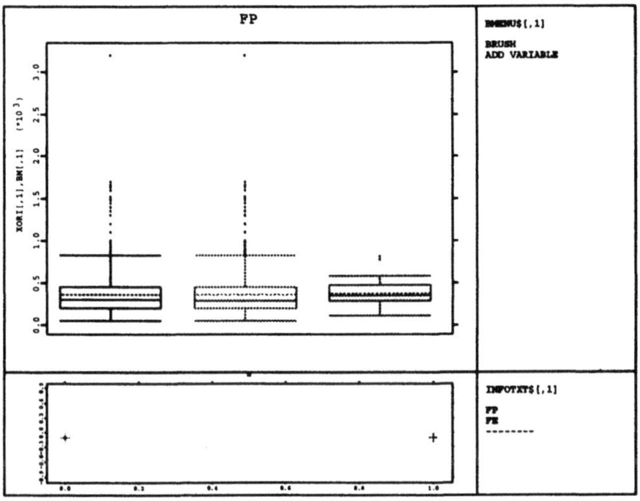

FIGURE 2.3. Subgroup of the variable FP by the variable FE. We see that in general the prices for flats in the east part of Berlin are higher (right boxplot) than in the west part of Berlin (middle boxplot). But in the west part of Berlin we have a larger range of prices. The boxplot for the whole dataset (left boxplot) is nearly the same as for the west part of Berlin. This is due to the fact that we have only 248 observations for the east part out of 14968.

Analyzing subgroups.

Often it is possible to decompose a dataset into subgroups, so we are interested to know how the distribution and outside values will behave on the subgroups. An example can be seen in Figure 2.3.

2.4 Quantile-Quantile Plot

Quantile-quantile plots are used to compare the distribution of random variables. Two types of comparisons are used:

1. to see if two random variables have the same distribution or
2. to compare a random variable with a predefined distribution.

For some statistical methods it is assumed that the distribution of the error has a special distribution. One of the most important coefficient to measure

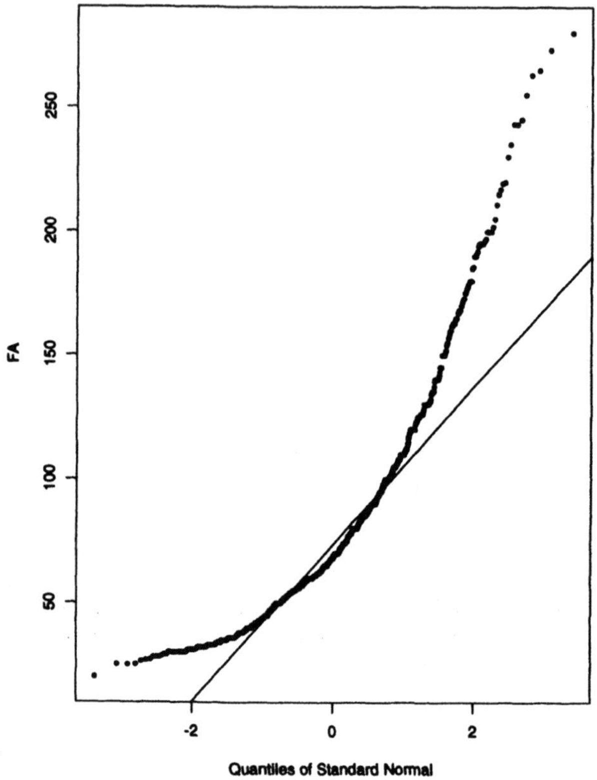

FIGURE 2.4. The quantile-quantile plot compares the variable FA with a gaussian distribution (see Figure 1.2). Since the datapoints deviate seriously from the line, it is clear that the assumption is not fulfilled that the variable FA is gaussian distributed.

the association between two continuous variables is the correlation coefficient:

$$r_{xy} = \frac{\sum_{i=1}^{n}(x_i - \bar{x})(y_i - \bar{y})}{\sqrt{\sum_{i=1}^{n}(x_i - \bar{x})^2 \sum_{i=1}^{n}(y_i - \bar{y})^2}}$$

The correlation coefficient can be computed for each dataset of two continuous variables, but we are also interested to test the coefficient at unequality of zero. To use the standard test, we need that both X and Y are normally distributed as assumption.

Here we can use the normal plot to explore if the distribution of X and Y is gaussian and to examine which datapoints are violating the normality.

It is obvious that such a plot is an exploratory tool. A test should be used (Kolmogorov-Smirnow, χ^2 or others) to indicate that the distribution is not gaussian.

2.5 Histograms, Regressograms and Charts

2.5.1 Histogram

Histograms are a graphical representation of the whole distribution of a dataset. In the time of text based terminals histograms were easily built up. The stem and leaf plot is a text based variation of a histogram, which allows to pick up some location parameters like quantiles, median etc.

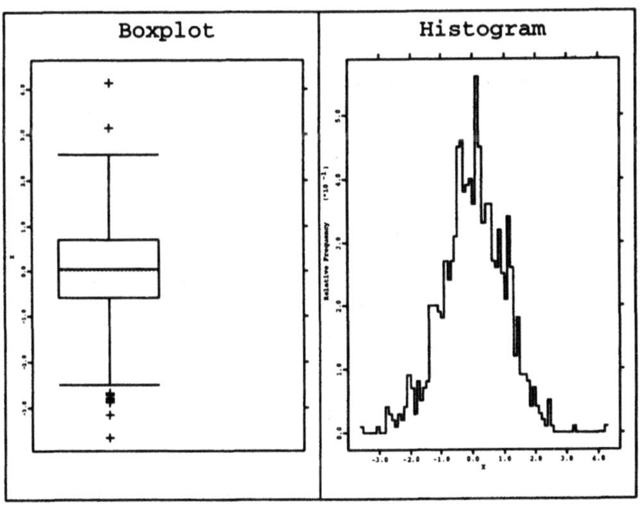

FIGURE 2.5. Boxplot and relative frequency histogram of 1000 data sampled from $N(0,1)$ (binwidth = 0.1, origin = 0).

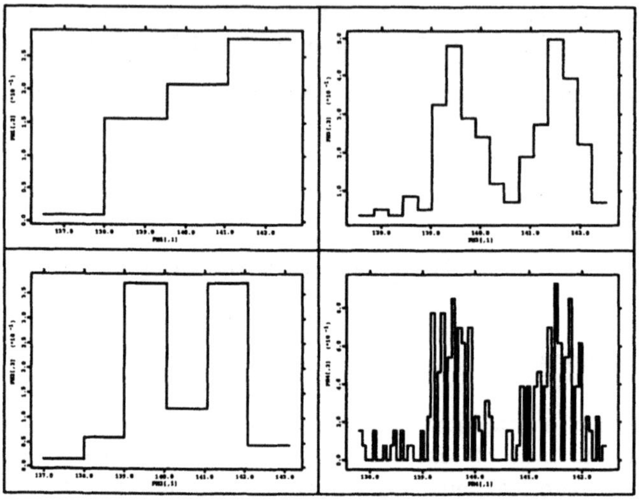

FIGURE 2.6. Histogram of the sixth variable of the Swiss banknote dataset. We see four histograms with a binwidth of 1.5, 1.0, 0.3 and 0.065. The bimodal structure is hidden if the binwidth is too big (left upper picture).

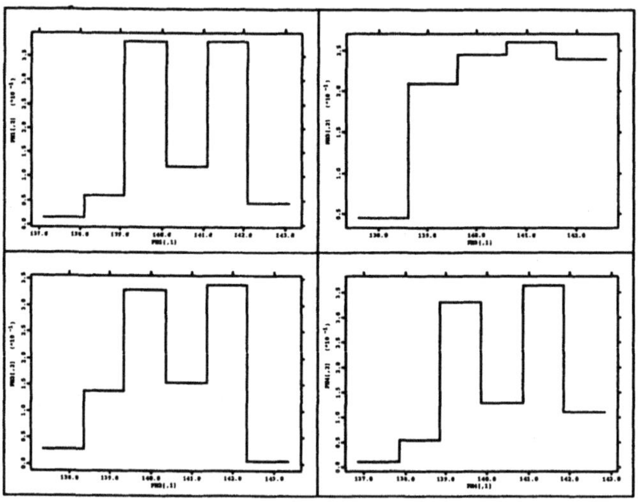

FIGURE 2.7. Histogram of the sixth variable of the Swiss banknote dataset. We see four histograms with a the same binwidth of 1.0, but the origin has different values 0.1, 0.35, 0.6 and 0.85. The bimodal structure is hidden in the right upper picture (origin is 0.6).

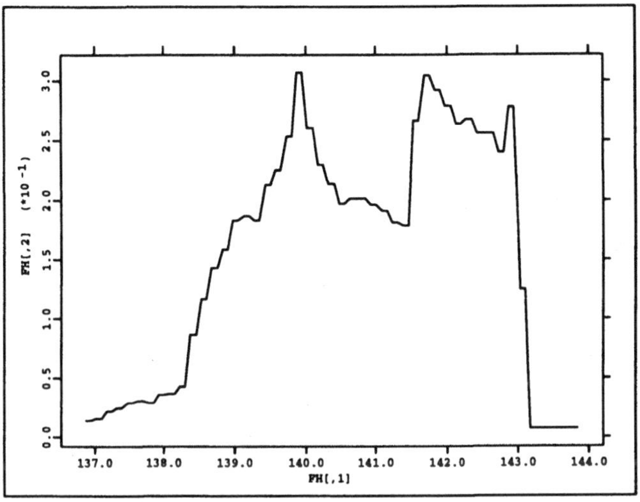

FIGURE 2.8. Averaged shifted histogram of the sixth variable of the Swiss banknote dataset. The histogram has the binwidth of 0.75 and is composed from 5 histograms with the origins 0.00, 0.15, 0.30, 0.45 and 0.60. The bimodal structure is now clearly visible and we could give some speculation if the structure is even trimodal.

A histogram provides more detailed informations than a boxplot as can be seen in Figure 2.5. It shows us how many data can be found at which location. Histograms appear in different forms, e.g. absolute frequency histogram, relative frequency histograms and other forms.

The construction of a histogram can be expressed as in Härdle (1991):

- Divide the real line into bins

$$B_j = [x_0 + (j-1)h, x_0 + jh) \quad j \text{ an integer}$$

  with binwidth h and origin x_0.

- Count how many data fall into each bin

- Depending on the wanted type of histogram we have to multiply with a constant. If the histogram represents a density function we write the histogram as

$$\hat{f}_{h,x_0}(x) = \frac{1}{nh}\sum_{i=1}^{n}\sum_{j} I(X_i \in B_j) I(x \in B_j)$$

Computationally it will be done in the following way:

1. Bin the data X_i and get the bincenters $\hat{B}_i$ and the bincounts C_i:

$$\begin{pmatrix} \hat{B}_1 & C_1 \\ \vdots & \vdots \\ \hat{B}_k & C_k \end{pmatrix}$$

2. Ensure that for empty bins a 0 will be inserted

3. Double your data:

$$\begin{pmatrix} \hat{B}_1 & C_1 \\ \hat{B}_1 & C_1 \\ \vdots & \vdots \\ \hat{B}_k & C_k \\ \hat{B}_k & C_k \end{pmatrix}$$

4. Add and subtract now the half of the binwidth δ:

$$\begin{pmatrix} \hat{B}_1 - \delta/2 & C_1 \\ \hat{B}_1 + \delta/2 & C_1 \\ \vdots & \vdots \\ \hat{B}_k - \delta/2 & C_k \\ \hat{B}_k + \delta/2 & C_k \end{pmatrix}$$

5. Plot this dataset

Two problems arise in connection with the drawing of histograms:

- How big should a bar (binwidth) of a histogram be?
- Where should we put the origin of the histogram?

To illustrate the problem see Figure 2.6 which shows the histogram for different bandwidths for a univariate dataset. In Figure 2.7 we can see what happens if we move the origin of the histogram.

The problem of the origin in the case of density estimation can be solved with average shifted histograms suggested by Scott (1985). The idea is to use a smaller binwidth and to distribute a observation over more than one bin. One practical solution would be to calculate the histogram for the different origins and to take x the mean of the values of the histogram at every point

$$\hat{f}_h(x) = \frac{1}{K}\sum_{i=1}^{K} \hat{f}_{h,x_0+ih/K}(x),$$

which is a specialized form given in Scott (1985). We can compare the Figure 2.8 with the Figure 2.7 and see that the problem of the origin is solved, but the problem of the binwidth remains. In the beginning of computational statistics it was suggested that the number of bins is proportional to $\sqrt{n}$. In `S-Plus` the number of bins is proportional to $\log_2(n)$. A calculation of the approximated mean integrated squared error (AMISE) of the histogram (Härdle 1991) leads to

$$AMISE(\hat{f}_{h,x_0}) = \frac{1}{nh} + \frac{h^2}{12} \parallel f' \parallel_2^2$$

with f' the derivative of the unknown density function and $\parallel . \parallel_2^2$ the squared L_2-norm of the density function. A minimization of the $AMISE$ leads to an "optimal" binwidth of

$$h_{min} \sim n^{-1/3}.$$

In practice this choice is not very helpful since we would have to know the unknown density function, but we could plug in a reference density for the unknown density function. Silverman (1986) suggested the gaussian distribution.

Another approach is to show a lot of different histograms and to change the binwidth interactively with a slider.

2.5.2 Regressogram

In the regressogram we try to estimate the unknown regression function by a stepwise constant function. The algorithm is similar to the algorithm for density estimation.

As an example see Figure 2.9 for a regressogram. Since the problems with them are mainly the same as for the histograms we will not go into details here. The regression estimator is very similar to the Nadaraya-Watson estimator.

2.5.3 Charts

Some charts for the plotting of noncontinuous variables have to be mentioned too. Barcharts and piecharts are used to represent such variables (see Figure 2.10 and Figure 2.11). Cleveland (1985) introduced dotcharts (see Figure 2.12) as an alternative to the pie and the barcharts.

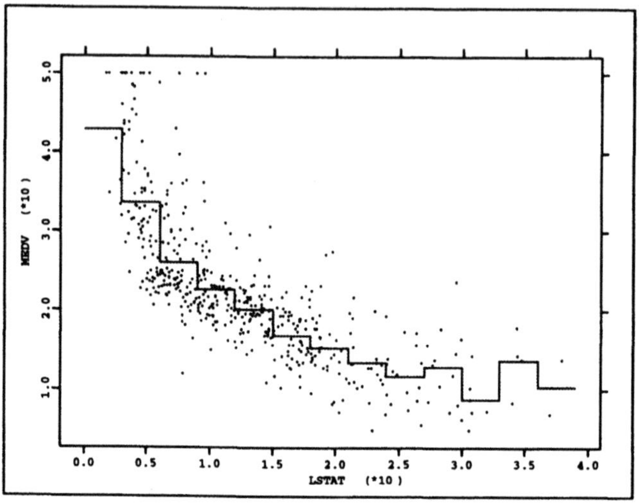

FIGURE 2.9. Regressogram of the variable LSTAT (percent lower status) and MEDV (median value of owner-occupied homes in thousand of dollars) of the Boston housing dataset. The binwidth is 3 and the origin 0.

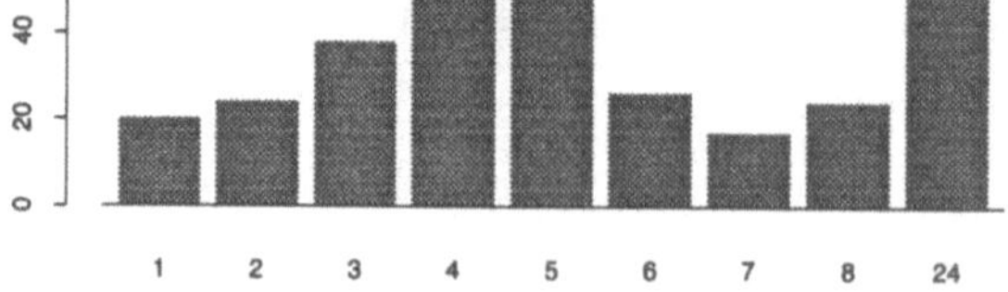

FIGURE 2.10. A barchart of the frequency of values of the variable exp(*RAD*) of the Boston housing dataset (accessibility to radial highways).

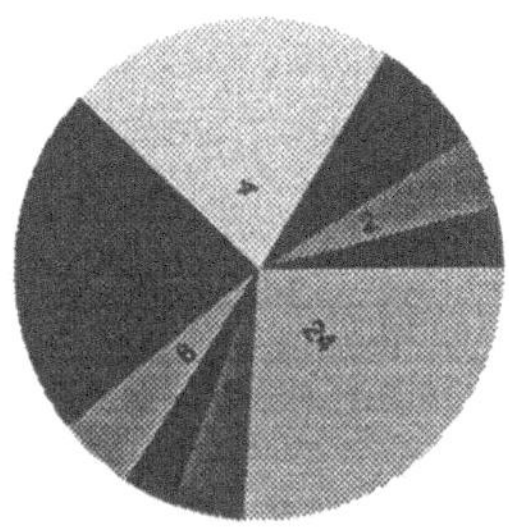

FIGURE 2.11. A piechart of the frequency of values of the variable exp(RAD) of the Boston housing dataset (accessibility to radial highways).

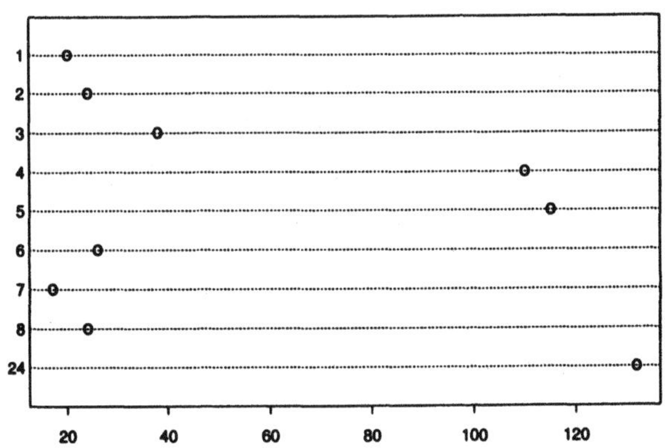

FIGURE 2.12. A dotchart of the frequency of values of the variable exp(RAD) of the Boston housing dataset (accessibility to radial highways).

2.6 Bivariate Plots

2.6.1 Scatterplot

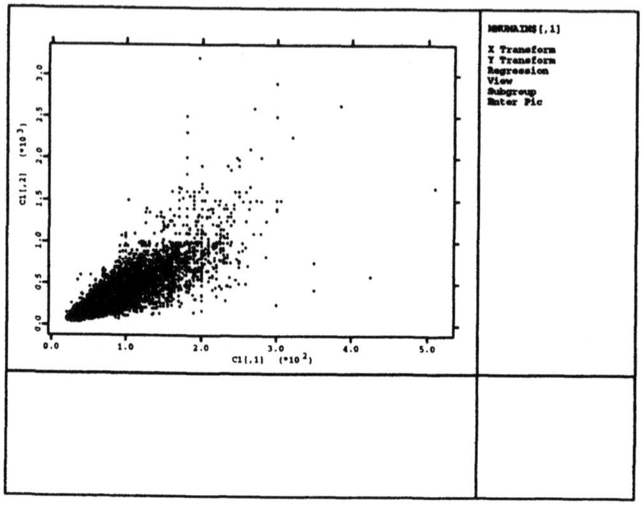

FIGURE 2.13. Scatterplot of the variables FA and FP of the whole Berlin flat data ($n = 14968$).

Aim.

One exploratory tool to analyze a two dimensional dataset is the scatterplot. Figure 2.13 tells us that we have a relationship between both variables.

A problem which occurs in a scatterplot is the "overplotting". It appears generally if we have too many datapoints as can be seen in Figure 2.13. To avoid that we can focus on some parts of the plot as can be seen in Figure 2.14. Here a rectangular brush was opened and we focus on the contents of the brush. What we need additionally is the possibility to jump back to a higher level of focussing and to move through the dataset and to zoom in again.

Subsets.

For further analysis of the data in Figure 2.13 we need the possibility to look at subsets of the data. Such a subset can be marked by different colours or by different forms. For this we need a brush which allows us to mark a datapoint or a subset of datapoints. The brush should be variable in size such that we

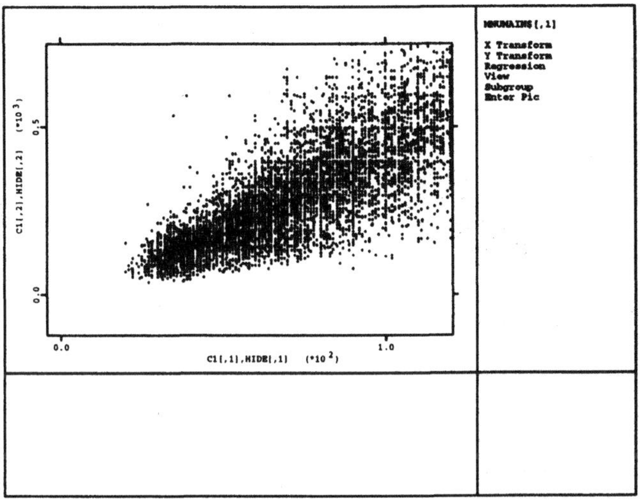

FIGURE 2.14. Scatterplot of the variables FA and FP of the whole Berlin flat data ($n = 14968$) focussed to the lower left corner of Figure 2.13.

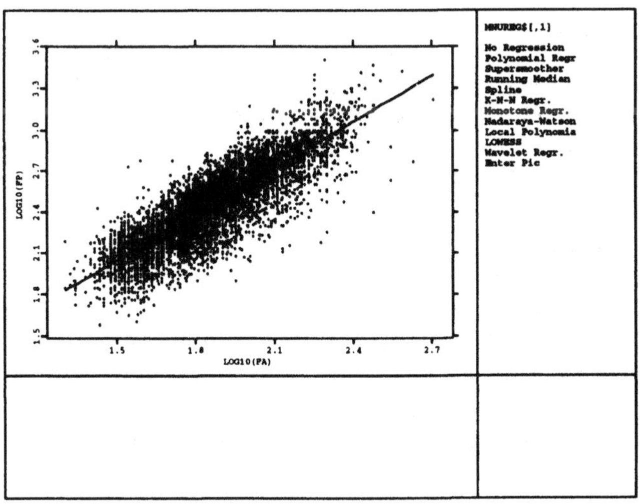

FIGURE 2.15. Scatterplot of the variables $\log_{10}(FA)$ and $\log_{10}(FP)$ of the whole Berlin flat data ($n = 14968$) and linear regression $\log_{10}(FP) = a \log_{10}(FA) + b$. This model of regression makes more sense since a retransformation of the variables leads to the model $FP = 10^b FA^a$ which reflects our expectation that the regression curve should run through $(0,0)$.

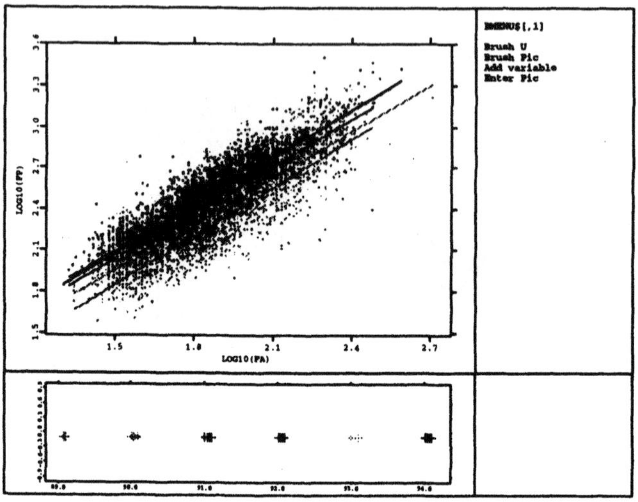

FIGURE 2.16. Scatterplot of the variables $\log_{10}(FA)$ and $\log_{10}(FP)$ of the whole Berlin flat data ($n = 14968$) and linear regressions on each year. We get the impression that from 1989 to 1992 the prices increased, dropped in 1993 and remained on the same level also in 1994.

can brush either one datapoint or a whole set of datapoints. Sometimes it might be useful to change the shape of the brush (see Figure 2.17), but it involves more complicated algorithms to figure out which datapoints are inside the brush. Even complicated areas can be marked easily if brush is changeable in size. For more details about brushing see section 2.7.

Transforming variables.

Sometimes we are not satisfied with the view we have on the data. From the modeling of the relationship in the Boston housing data (see Table A.1) we know that for example the variable LMV is the logarithm of the median price so we would like to transform the variables in our example. In fact Figure 2.15 shows a more interesting behaviour than Figure 2.13.

Regression.

We often have the problem to show datasets which are linked, e.g. a dataset and its regression function, so we need to put more than one dataset into a scatterplot (see for example Figure 2.15). The brushing of a subset and the representation of the corresponding regression lead to an interesting view to the data (see Figure 2.16).

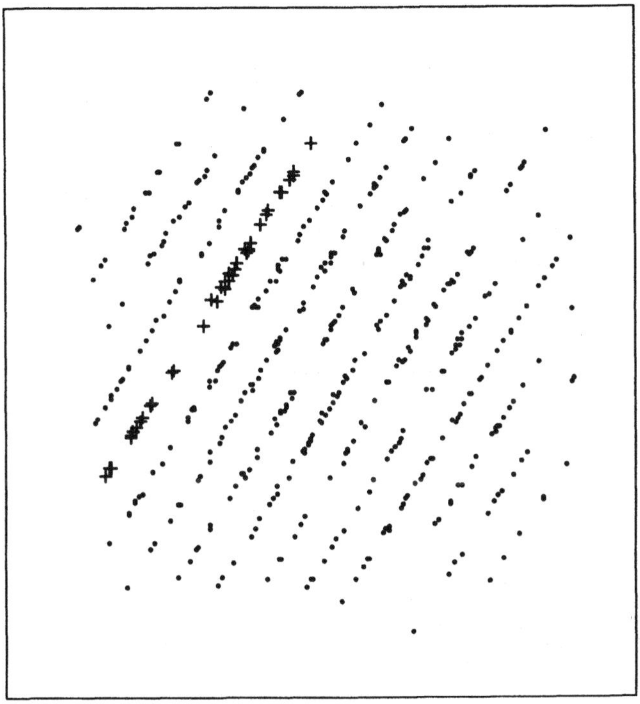

FIGURE 2.17. Two dimensional projection of RANDU data in XGobi. A brush was used to mark one plane of data points.

Control elements.

As statisticians we need to control a lot of features in such a plot. If we use lines in the plot then the datapoints have to be connected by lines of different colour, thickness and type. Many texts have to be checked: The title of a window, text at the position of the datapoints etc, the plotting of (multiple) axes including scaling, tick marks, origin and text at the tick marks. For time series the text at the tick marks has another format than in a 2D-scatterplot, sometimes it is necessary to add datapoints, texts and lines. For example in a contour plot it would be nice to know at which contour he have cutted. These features point at the capabilities of drawing programs.

The scaling of a plot means how to choose the sizes of the axes in the beginning. In general we will choose the scale in a way that we can see the whole dataset. Later the user may zoom in and out of the dataset. In the case that we are plotting lines it is appropriate to choose the median of the slopes to be ± 1 as suggested in Cleveland, McGill & McGill (1986). A fitting of the axes to the data might hide obvious features.

2.6.2 Sunflower Plot

A problem mentioned earlier is the overplotting. If we distribute one dot of ink on a paper for every datapoint having to plot a lot of datapoints we get a big spot where we cannot see anything. One solution would be to reduce the size of the plotsymbol. But in the sunflower plot binning is used. We plot different symbols depending on how many datapoints fall into a bin (see Figure 2.18). Of course we can use colours (image plots), the brightness or sizes of the plot symbol.

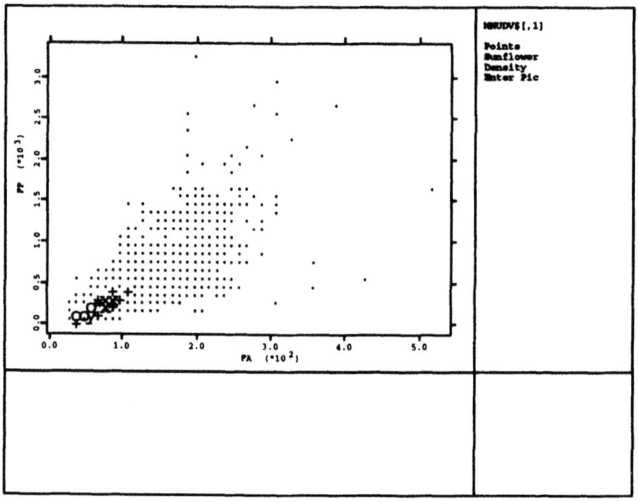

FIGURE 2.18. Sunflower plot of the variables FA and FP of the whole Berlin flat data ($n = 14968$, $X_{bin} = 10\,m^2$, $Y_{bin} = 100.000\,DM$). In contrast to Figure 2.13 we can see that most flats have less than $100\,m^2$ and are cheaper than $500.000\,DM$.

2.6.3 Other Views

Of course we also need other bivariate plots than scatterplots. For example it can be difficult to overview a density function depending on two random variables which can be plotted in a 3D-scatterplot. To get a better overview we can use contour plots as in Figure 2.19. The fixing of the contour lines

$$f(x, y) = c_i \quad i = 1, 2, \ldots$$

has to be done interactively if we do not have a special aim in mind, e.g. seeing the global maximum. Another possibility are image plots. Here the

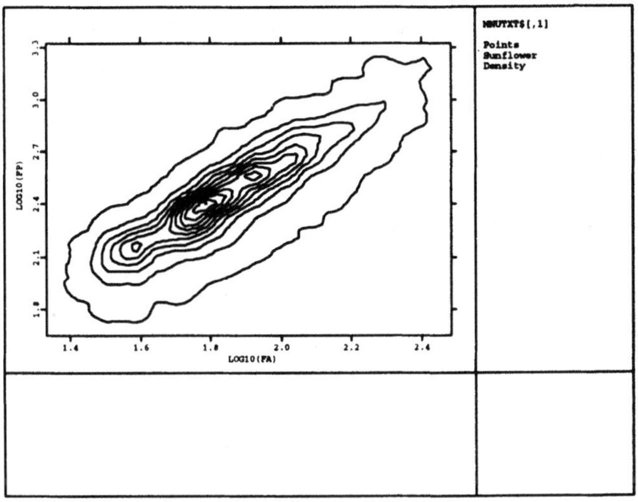

FIGURE 2.19. Scatterplot of the variables $\log_{10}(FA)$ and $\log_{10}(FP)$ of the whole Berlin flat data ($n = 14968$). A density estimate over the data was done with $h_x = h_y = 0.01$. A further investigation would show that the peeks we see in the contour plot represent a certain number of rooms.

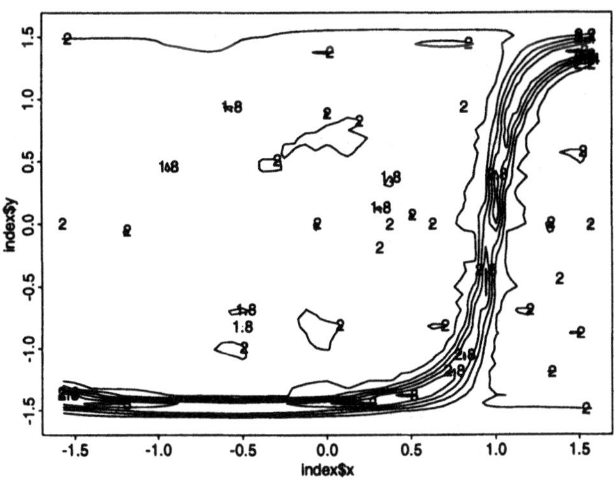

FIGURE 2.20. Contour plot of a projection pursuit index function. The index is the Friedman-Tukey index (see chapter 3) with the triweight kernel and a bandwidth $h = 0.05$ on the RANDU data (see also Figure 2.17).

colour of a small rectangular area represents the actual coordinate in the z-axis. In the case of the image plots the question of the choice of colours and therefore the question of the colour models used will appear.

2.6.4 Trees

Trees are used for visualization in several statistical techniques,

- as dendrograms in hierarchical cluster analysis (see section 3.1)
- in classification and regression trees.

Dendrograms are used to decide how many clusters are available in the data. In classification and regression trees the spliting rules are visualized (Breiman, Friedman, Olshen & Stone 1984).

2.7 Scatterplot Matrices

The scatterplot matrix consists of a set of scatterplots. It is a tool to analyse a multivariate dataset. For each pair of variables of a multivariate dataset we produce exactly one scatterplot as shown in Figure 2.21.

We can reduce the amount of scatterplots by excluding the upper right half of the scatterplots as it is just a mirrored view of the scatterplots in the left lower half. If we drop these pictures we get Figure 2.22.

The real power of a scatterplot matrix will we get by brushing. Normally the brush is a rectangular area which gives a specific colour and form to all datapoints in the brush area which can be chosen by the user. It is assumed that if the i-th datapoint in one scatterplot changes its colour and/or form then the i-th datapoint in all other scatterplots will change accordingly.

The facilities of a brush (Fisherkeller et al. 1988) should be:

- Deleting
  All datapoints in the brush are masked out. Often the colour of a datapoint is set to the background colour.
- Highlighting
  A datapoint gets another colour and/or form. The user can choose them interactively.
- Transient/nontransient
  If a datapoint has changed the colour and/or the form and the brush is

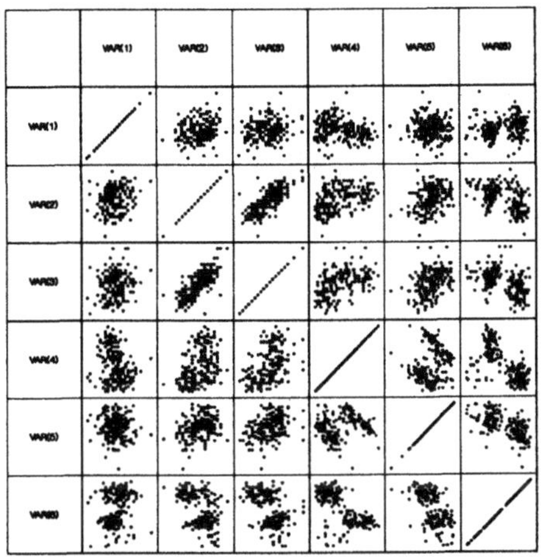

FIGURE 2.21. Scatterplot matrix of the six dimensional Swiss banknote dataset (from SYSTAT).

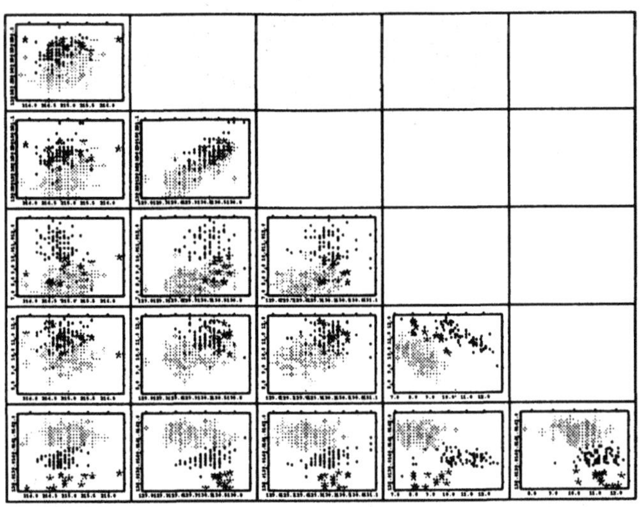

FIGURE 2.22. Scatterplot matrix of the six dimensional Swiss banknote dataset. The three clusters visible in the second window from the left in the last row are brushed with 3 different colours and symbols.

moved further, then we have two possibilities if the datapoint falls out of the brush. It either keeps the new colour and/or the new form or it returns to the colour and/or the form before brushing.

- Brush shape
  The brush shape should be variable. Lasso functions will be useful to brush nonrectangular areas, but they will be difficult to implement. At least a brush should offer different sizes so that we are able to brush complicated areas.

2.8 Three Dimensional Plots

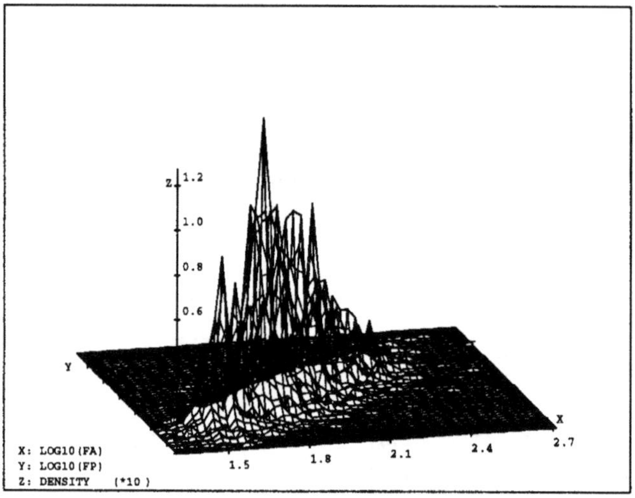

FIGURE 2.23. 3D-plot of a bivariate density estimate of the variables $\log_{10}(FA)$ and $\log_{10}(FP)$. As in Figure 2.19 we can see the peeks.

Aim.

As a scatterplot is a tool to analyze the relationship between two variables a 3D-scatterplot is a tool to analyze the relationship between three variables. Since our output tool is still a screen we need additional techniques to give the eye the impression that we looking at something three dimensional.

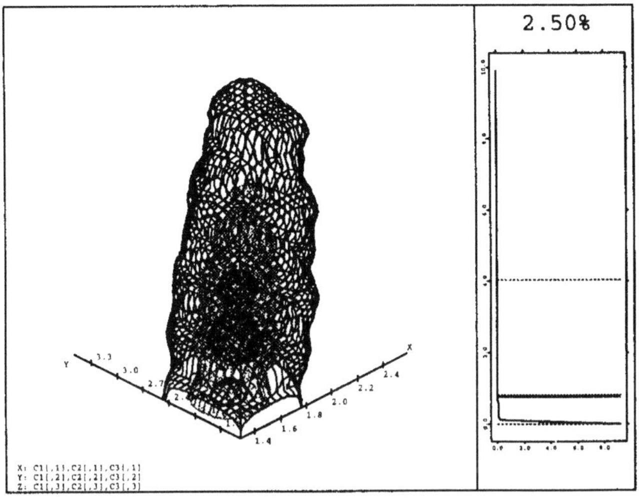

FIGURE 2.24. 3D-plot of a trivariate kernel density estimate of the variables $\log_{10}(FA)$, $\log_{10}(FP)$ and FR (bandwidth $h = (0.23, 0.32, 0.75)$). As expected we have a clear relationship between these variables. If they would be uncorrelated the density estimate would look more like unit balls. Instead of different colours we have used the gray scale representation of them.

Spinning.

The first attempt has been done with spinning which means we rotate the dataset parallel to one of the screen coordinate axes (see e.g. **MacSpin**). A lot of problems has risen in this context. The first one is the internal representation of a three dimensional dataset such that a rotation will appear to the eye as rotation and not as a set of blinking pictures. Models have been developed to represent the continuous data internally on an integer grid and to executed the rotation on this grid. Fortunately the numerical and graphical power of computers has improved so much that this is no longer a problem.

Sizing.

Another possibility to get a three dimensional effect is to draw the datapoints which are closer to the observer thicker than those being far away. If we rotate this dataset we need additional computational effort to compute the distance from the observer. We also loose the possibility to supply a datapoint with a form of arbitrary size.

Stereoplot.

Another approach is to split a datapoint into two datapoints which have a small distance from each other. If we colour one datapoint red and the other one green and if we use red-green-glasses we will get a three dimensional picture of our dataset. The disadvantage is of course that we had to double the number of observations and that we always needed red-green-glasses.

Rocking.

A much more interesting technique seems to be the "rocking" of a dataset. If we look at a three dimensional scatterplot the picture does not stand still but moves between two position by rotation. Since datapoints being more distant will move by greater distances than closer observations we are able to recognize how far away the observations is compared to the other datapoints.

The advantage is that we only have to compute two different positions for that moment when we stop the rotation. The computational effort is not too big and the routines for the rotation are already available.

Surface.

3D-scatterplots are not only used to show datapoints. They are also used to show different kinds of surfaces (see Figure 2.23 and Figure 2.24).

For the trivariate kernel density estimate in Figure 2.24 an interactive choice is necessary of the levels c_{red}, c_{green} and c_{blue} to plot the contours of

$$\hat{f}(x_i, y_j, z_k) = c_{\text{colour}}.$$

Colour models.

Since we are using colours we have to choose between different colour models. As each model uses a different basis to compose a colour, each has its own advantages and disadvantages:

- RGB
  RGB (red-green-blue) is the most commomly used colour model. Our TV pictures on the screen use this model. Every colour is composed of a partition of red, green and blue. We have a lot of knowledge available about the eye's sensitivity to RGB-triplets. Every window system provides an RGB-triplet for composing a colour. A problem appears if someone has to compose a colour by himself as some experience is needed.

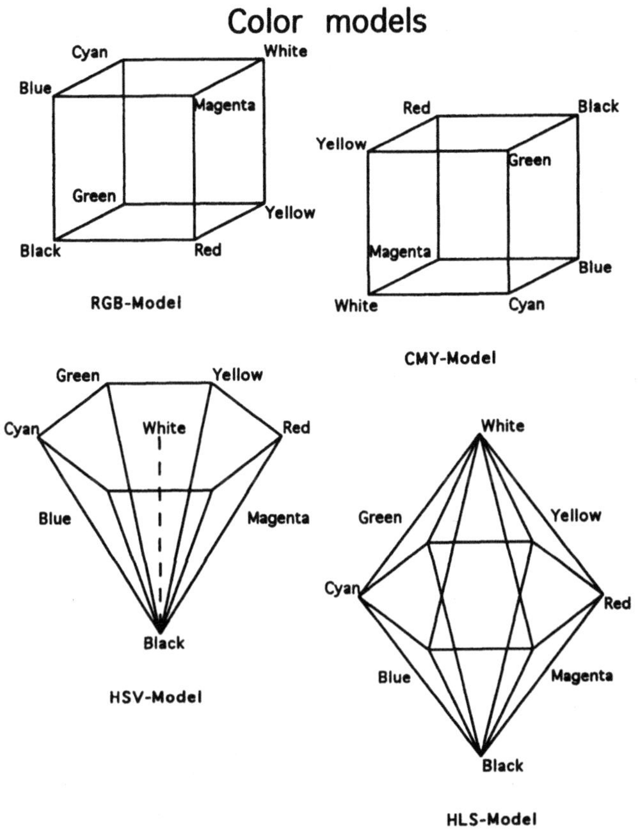

FIGURE 2.25. Representation of colour models

- YIQ
  The YIQ model was designed for transmission efficiency in colour broadcast TV. It can be calculated from the RGB-model by

$$\begin{pmatrix} Y \\ I \\ Q \end{pmatrix} = \begin{pmatrix} 0.30 & 0.59 & 0.11 \\ 0.60 & -0.28 & -0.32 \\ 0.21 & -0.52 & 0.31 \end{pmatrix} \begin{pmatrix} R \\ G \\ B \end{pmatrix}$$

- CMY
  The CMY (cyan-magenta-yellow) is widely used for colour printing devices. It can be calculated easily from the RGB-model by

$$\begin{pmatrix} C \\ M \\ Y \end{pmatrix} = \begin{pmatrix} 1-R \\ 1-G \\ 1-B \end{pmatrix}$$

- HSV/HLS
  The HSV- (hue-saturation-value) and the HLS-model (hue-lightness-saturation) are designed for a user-friendly composition of colours. The hue distinguishes between different colours like red, yellow, green, cyan, blue and magenta. The saturation describes how little the colour is diluted with white, e.g. pink and red, sky blue and royal blue.... The lightness or value describes the intensity of a colour. Both systems can be represented by a single- or double-hexacone.

- Munsell system
  None of the described models consider the sensitivity of the human eye. In computer systems the RGB model is composed by three integers which have a range of 255 ($256^3 = 16.7$ Mio.). But can a human eye really distinguish the colour $(0,0,0)$ from $(1,1,1)$? So Munsell built up a scale such that we have equally perceived distances in colour space. This scale is subjective, but it is based upon the evaluation of many observers.

The statistical importance of colour scale appears in contour plots in the three dimensional case (Scott 1992) and in image plots in the two dimensional case. The three most interesting colour-models, RGB, HSV and HLS can easily be implemented in statistical software. The Munsell system is based on huge tables so that an implementation is only done if necessary.

A problem that often arises in (statistical) programs is how to transfer the background colour of the screen (mostly black) to the background colour of the printer (mostly white). An easy exchange of black and white is not possible, because if someone uses a gray scale starting with white and ending with black, e.g. in a contour plot or in an image plot, the exchange would destroy the whole palette. To solve such a problem the HLS system can be used. The RGB-colour will be translated to HLS-colour and the saturation s will be set to $1-s$. That ensures that colour with $s = 0.5$ will not change the RGB-colour. Additionally the light colours which are a strong contrast to a dark background will become dark colours on the printer, which also will be a strong contrast on the paper.

2.9 Higher Dimensional Plots

2.9.1 Three Dimensional Scatterplots with Colour, Form and Size

The simplest idea to show multidimensional data is to use a 3D-scatterplot and to use different colours, forms and sizes for the datapoints to indicate additional dimensions. But the number of sizes and forms is limited (the number of colours sometimes too), so if a variable is continuous we are not able to see this continuity. This kind of plots can be used if the variable only has a small number of discrete values.

2.9.2 Chernoff Faces

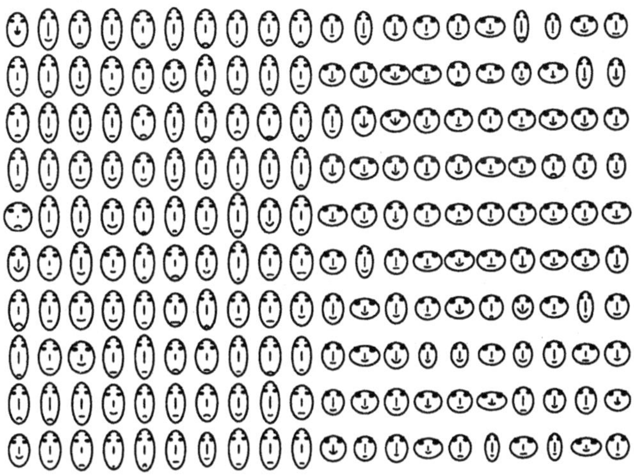

FIGURE 2.26. Chernoff faces constructed with the algorithm of Chernoff. The observations of the Swiss banknote dataset are coded in face parts: variable 1 is the width of the mouth, variable 2 the curvature of the mouth, variable 3 the location of the mouth, variable 4 the shape of the face, variable 5 the length of the nose and variable 6 the area of the face. All 200 banknotes are displayed. We can easily see that we have 2 different types of banknotes.

Chernoff faces and other glyphs (star diagram, trees etc.) are also used to represent multivariate data. Chernoff (1973) introduced the faces in statistics. He has coded 15 variables in different faces parts (see Figure 2.26). Flury & Riedwyl (1981) were not satisfied with the look of the faces if the data have extreme values. They stated that the faces do not look anylonger like faces, and as a consequence an observer will be more attracted by these nonhuman looking faces than by the human-looking ones. Additionally they coded 36 variables and developed a face that consists entirely of polygons (see Figure 2.27).

FIGURE 2.27. Chernoff faces constructed with the algorithm of Flury and Riedwyl. The observations of the Swiss banknote dataset are coded in face parts: variable 1 is the nose line, variable 3 the curvature of the eyebrow, variable 4 the eye size and the size of the mouth and variable 6 the density of the eyebrow, the face line and the darkness of the hair. We see that observation 70 of the dataset looks very different to the others.

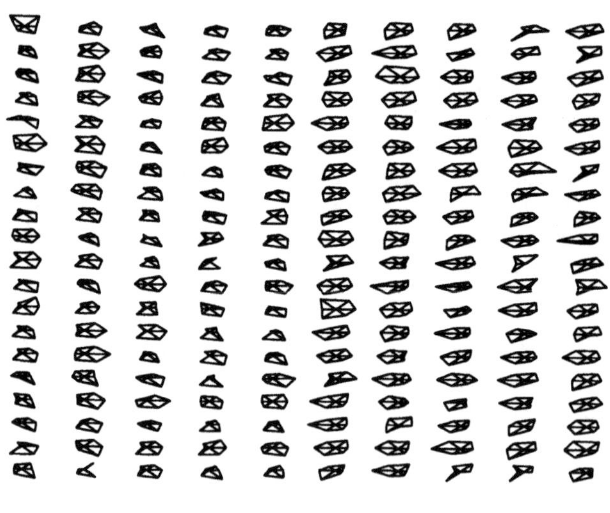

FIGURE 2.28. Star diagrams constructed from the Swiss banknote dataset. We can see that stars on the right look fatter.

The problem with the Chernoff faces is that we do not perceive all face parts equally. For example the size of the pupil will not receive the same attention as the darkness of the hair.

If we use the star diagram for all variables then they will be perceived equally. Since Chernoff faces overemphasize some variables we are able to see differences well. This is no longer the case with star diagrams. Here only extreme values will lead to a bigger attention (see Figure 2.28).

In the case of Chernoff faces a tool offering such glyphs has to allow an easy reassociation with the variables of the face parts. It holds for all glyphs that it should be possible to sort the data for a certain variable. This will lead to faster recognition if a structure is connected to a certain variable.

2.9.3 Parallel Coordinate Plot

If we try to represent multivariate data we will have a loss of information. In the grand tour or Andrews curves orthogonal projections of the data are used. Inselberg (1985) and Wegman (1990) tried to go another way. He gave up the idea of orthogonality and put all coordinate axes parallel to all others. Each datapoint can be marked on the axes and we can draw lines which are connecting observations through axes (see Figure 2.29).

This can be seen as a projective transformation. The hope is that the geometric information from standard euclidean space which carries statistical information is mapped in geometric structures in parallel coordinate space. One statistical information which carries over is the correlation between variables as can be seen in Figure 2.30.

Another structure that carries over is the presence of clusters as can be seen in Figure 2.31. If the clusters are separable in one or more dimension this can be recognized in the parallel coordinate plot.

A drawback of parallel coordinate plots is that we can not overview them if the dataset becomes large. A solution to this problem is to draw line densities $f(x, q)$ on the lines parallel to the coordinate axes ($q \in [0, 1]$). An example can be seen in Figure 2.32.

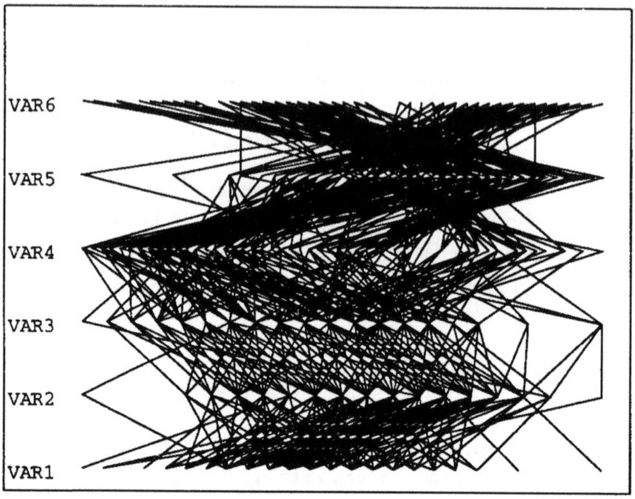

FIGURE 2.29. Swiss banknote dataset in the original version of the parallel coordinate plot. All 200 observations are shown and the variables are rescaled on [0, 1].

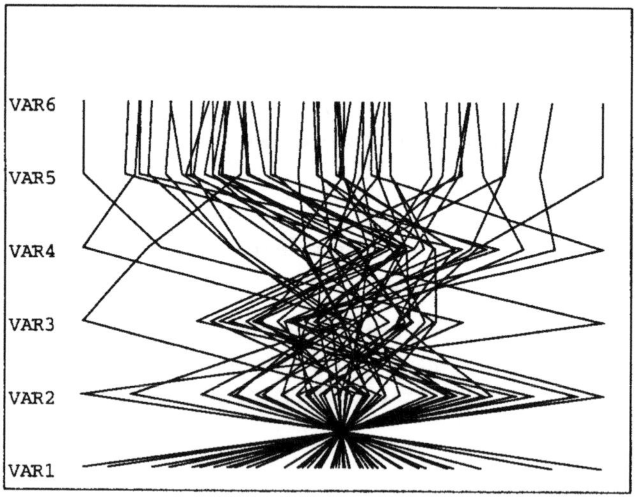

FIGURE 2.30. Simulated dataset with different values for the correlation between variables. $corr(X_1, X_2) \sim -1$, $corr(X_2, X_3) \sim -0.5$, $corr(X_3, X_4) \sim 0$, $corr(X_4, X_5) \sim 0.5$, $corr(X_5, X_6) \sim 1$. We can see that the lines become less wild as the absolute value of the correlation increases.

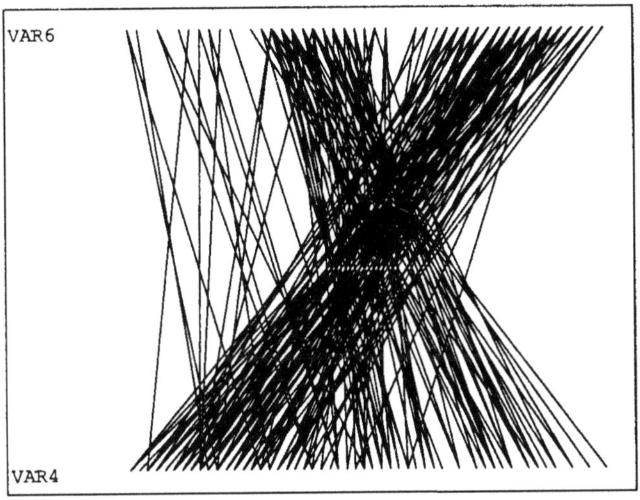

FIGURE 2.31. Parallel coordinate plot of the fourth and the sixth variable of the Swiss banknote dataset (see also Figure 2.21).

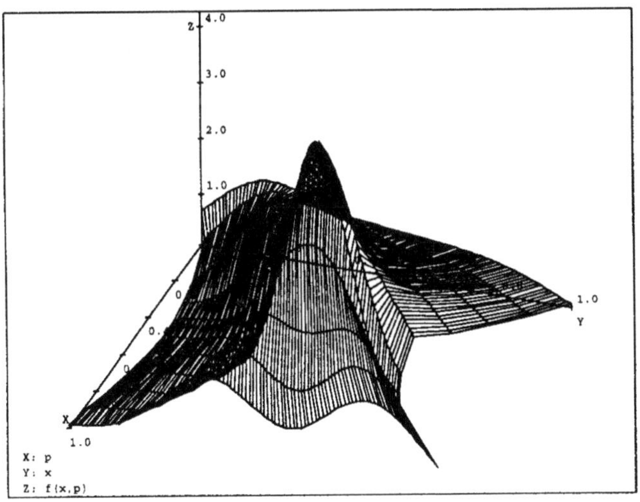

FIGURE 2.32. Parallel coordinate plot with line densities of the Swiss banknote data set. The plot deviates from a standard normal distribution.

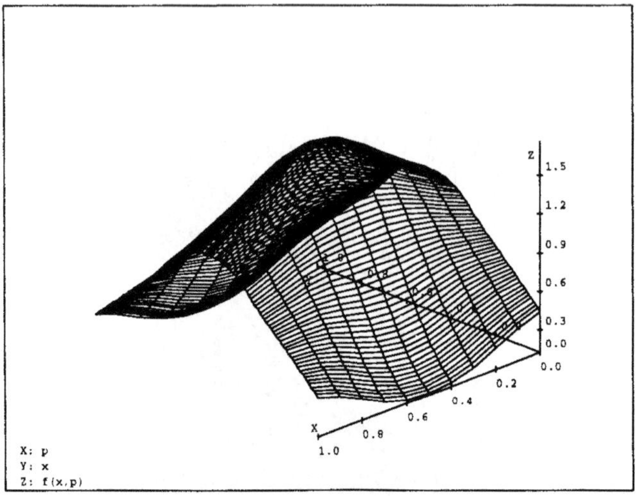

FIGURE 2.33. Parallel coordinate plot with line densities of the two gaussian random variables with mean zero and the covariance matrix being the unit matrix.

If we know the density of two random variables it is possible to calculate the intermediate density $f(x, p)$ from the formula

$$V(p) = (1-p)X_1 + pX_2$$

for a two dimensional dataset. Especially if we have two gaussian distributions such that $X = (X_1, X_2)$ with $X_1 \sim N(\mu_1, \sigma_1)$, $X_2 \sim N(\mu_2, \sigma_2)$ and $corr(X_1, X_2) = \rho$ the intermediate density $f(x, p)$ will also be gaussian with

$$\mu_p = (1-p)\mu_1 + p\mu_2 \quad (2.1)$$

$$\sigma_p^2 = (1-p)^2\sigma_1^2 + p^2\sigma_2^2 + 2p(1-p)\rho\sigma_1\sigma_2. \quad (2.2)$$

An example with $\mu_1 = \mu_2 = \rho = 0$ and $\sigma_1 = \sigma_2 = 1$ can be seen in Figure 2.33. Instead of 3D-surface plots contour plots and image plots can be used.

2.10 Basic Properties for Graphical Windows

In the preceeding sections we have examined a lot of graphics which are used in statistical computing. We can divide the graphics into two basic classes:

1. graphics which consist of drawing datapoints, lines and text and
2. graphics required for much more complex objects.

Piecharts, Chernoff faces and glyphs in general belong to the second class of graphics while all other graphics belong to the first class.

For the implementation in a statistical software we need at least two kinds of graphical windows: one that will give us special graphics from the second class, and another for the specific demands of drawing graphics of the first class:

- Datapoints in various colours, sizes and forms. Here forms also include strings and small rectangles.
- Lines of various colours, types and thickness.
- In future we might also need to draw complete areas, so the data structures for graphical objects would have to consider that.

In the regression figure we have seen that the graphics need to be manipulated interactively or by command from a program.

3

Some Statistical Applications

Summary

Here some applications are discussed (cluster analysis, teachware, regression methods). The cluster analysis will serve as an example for the use of graphics. Teachware needs to be highly interactive and we shortly discuss the approach of Proenca (1995). The section about regression methods shows how detailed a programming language should be. The trade-off between speed and understanding in a statistical routine still plays an important role.

3.1 Cluster Analysis

3.1.1 Introduction

Cluster analysis algorithms are tools used frequently in ecology, biological science, marketing, chemistry, geology, social science, economics, archaelogy, ornithology etc. Cluster analysis attempts to detect structure in the data or at least provides to reduce the number of the observations. This technique divides a set of points into a subset in such a manner that similar points belong to the same cluster, whereas dissimilar ones are allocated into different clusters.

The cluster analysis consists of two different kinds of methods

- hierarchical methods and
- partitioning methods.

The partitioning methods require an initial classification. This means that the number of clusters is fixed. They try to exchange observations between the clusters to improve some criteria of goodness. The algorithms can be divided into two classes: the iterative ones and the noniterative ones. In the noniterative algorithms an observation can be classified only once and can not be exchanged afterwards to another cluster. The main advantage of these algorithms is that they reduce the amount of computing considerably. The iterative algorithms do allow a reclassification more than once.

The hierarchical methods do not need an initial classification, and the number of cluster needs not to be known previously. But these algorithms need the distances between the datapoints. The distance metric has been chosen by the user. Again the algorithms consist of two different methods. Agglomerative algorithms start with n clusters so that every datapoint is represented by one cluster. Now iteratively the algorithm will merge clusters together until the whole dataset consists of one cluster. The other method, the divisive algorithms, go exactly the other way. They start with one cluster and try to decompose the dataset into two subclusters. The process is iterated on the subclusters until each cluster consists of one datapoint.

It is useful to combine both methods so that we first execute a hierarchical cluster analysis to choose the number of clusters and then a partitioning method to improve our result.

In the following subsection we will restrict to one partitioning algorithm and one hierarchical algorithm. For further details on other algorithms see Mucha (1992a).

3.1.2 The k-means Algorithm

The k-means algorithm was developed by Hartigan (1975). As mentioned we need an initial classification. Possible choices are a random generated classification, prior knowledge or categorization of the first principal component. The k-means algorithm now tries to minimize the sum of the within cluster variances

$$V_K = \sum_{k=1}^{K} \sum_{i=1}^{n} \delta_{i,k} m_i d_Q^2(x_i, \bar{x}_k)$$

with K the number of clusters, n the number of observations, $\delta_{i,k}$ an indicator function, which is 1 if the i-th point is in the k-th cluster and 0 if not, and m_i a weight for the observation i.

$$d_Q^2(x, y) = (x - y)^T Q(x - y)$$

represents a weighted squared euclidean distance with a weight matrix Q. The weights normally describe the weight of variables, and classical choices are

- the trivial weights $Q = I$ or

- the standard weights $Q = diag(1/s_{j,j})$ with

$$s_{j,j} = \frac{\sum_{i=1}^{n} m_i (x_{i,j} - \tilde{x}_j)^2}{\sum_{i=1}^{n} m_i}$$

$$\tilde{x}_j = \frac{\sum_{i=1}^{n} m_i x_{i,j}}{\sum_{i=1}^{n} m_i}$$

Obviously other distances can be used too, but the weighted squared euclidean is the most common one.

The algorithm has to incorporate an optimization algorithm on a stepwise function which appears to be a difficult task.

An improvement for the partitioning algorithm can be done by the use of adaptive weights. We repeat the partitioning algorithm until the computed partition or the adaptive weights change no longer. In the first step we use the standard weights and in each following step we compute pooled standard deviations

$$sp_j = \frac{\sum_{k=1}^{K} \sum_{i=1}^{n} \delta_{i,k} m_i (x_{i,j} - \bar{x}_{k,j})^2}{\sum_{i=1}^{n} n m_i}$$

with

$$\bar{x}_{kj} = \frac{\sum_{i=1}^{n} \delta_{i,k} m_i x_{i,j}}{\sum_{i=1}^{n} \delta_{i,k} m_i}.$$

The inverse of these pooled variances will be plugged in as weights for the variable in the next step. Following Mucha (1992b) and Mucha & Klinke (1993) it seems that these methods are a little more intelligent than the standard k-means algorithms.

3.1.3 The Agglomerative Algorithm

All agglomerative methods follow the same scheme:

1. Find the minimal distance $d(i, j)$ between two clusters
2. Merge the two clusters to a cluster k
3. Compute the new distances of each cluster

   $$d(k, l) = \alpha_i d(i, l) + \alpha_j d(j, l) + \beta d(i, j) + \gamma \mid d(i, l) - d(j, l) \mid$$

   For the value of the constants for the different methods see Table 3.1.
4. Go to 1 until we have more than one cluster

For the agglomerative methods a choice of the distance is important. First we have to distinguish between continuous and noncontinuous variables. For the continuous case we have a lot of distances available. If we assume that $X = (x_{i,j})_{i=1,\dots,n;j=1,\dots,p}$ then we have the following distances

- the euclidean distance:

$$d(i, j) = \sqrt{\sum_{k=1}^{p}(x_{i,k} - x_{j,k})^2}$$

- the Manhattan distance:

$$d(i, j) = \sum_{k=1} \mid x_{i,k} - x_{j,k} \mid$$

- the maximum distance:

$$d(i, j) = \max_k \mid x_{i,k} - x_{j,k} \mid$$

- the cosine distance:

$$d(i, j) = \sqrt{2 - \frac{2\sum_{l=1}^{p} x_{i,l}x_{j,l}}{\sqrt{\sum_{l=1}^{p} x_{i,l}^2 \sum_{l=1}^{p} x_{j,l}^2}}}$$

- the χ^2 distance:

$$d(i,j) = \sum_{k=1}^{p} \frac{1}{\sum_{l=1}^{n} x_{l,k}} \left(\frac{x_{i,k}}{\sum_{l=1}^{p} x_{i,l}} - \frac{x_{j,k}}{\sum_{l=1}^{p} x_{j,l}} \right)$$

The choice of the distance depends on the problem we have. If we want to cluster airports with the aim of minimizing the traveling costs, we will use the euclidean distance since airplanes can go straight from one airport to another. But if we want to cluster supermarkets of a company to minimize transportation costs for goods, we will use the Manhattan distance, since streets often follow a rectangular mesh.

In contrast to the measures for continuous variables we also have measures for noncontinuous variables. The distances given below are for binary variables, but for other input data we can categorize the variables by

$$cat_v(x_i) = \begin{cases} 0 & \text{if } x_i < v, \\ 1 & \text{otherwise.} \end{cases}$$

For discrete variables with more than 2 values the variable can be decomposed into several variables in the form of

$$dec_j(x) = \begin{cases} 1 & \text{if } x_i = j, \\ 0 & \text{otherwise.} \end{cases}$$

Thus it is sufficient to give distances for binary variables:

- the Tanimoto distance:

$$d(i,j) = 1 - \frac{\#11 + \#00}{\#11 + \#00 + 2(\#10 + \#01)}$$

- the Jaccard distance:

$$d(i,j) = 1 - \frac{\#11}{\#11 + \#01 + \#10}$$

- the matching distance:

$$d(i,j) = 1 - \frac{\#11 + \#00}{p}$$

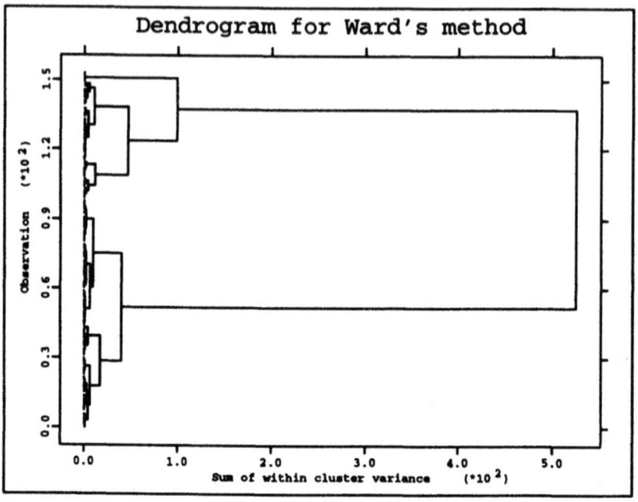

FIGURE 3.1. Dendrogram for the Swiss banknote data. On the x-axis we see how the sum of within cluster variance increases if we merge more and more clusters. On the y-axis we see the observations. A good choice for cluster would result in 2, 3 or 5 clusters. In fact in Figure 9.10 on page 186 in Polzehl & Klinke (1995) we can already see three distinguishable clusters.

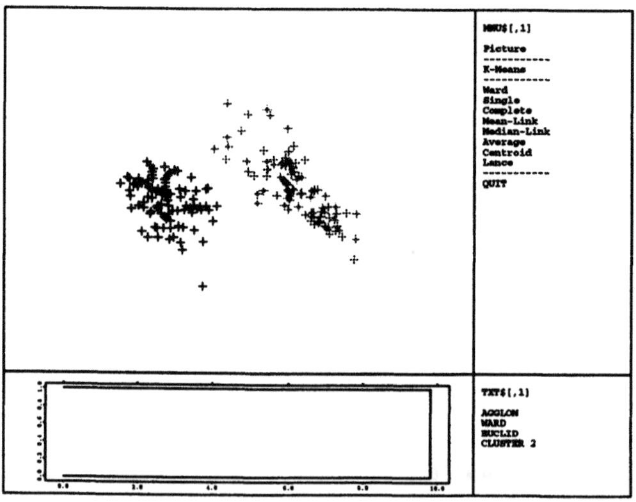

FIGURE 3.2. Principal component plot of Swiss banknote dataset. With the Ward method and the euclidean distance we have chosen 2 clusters.

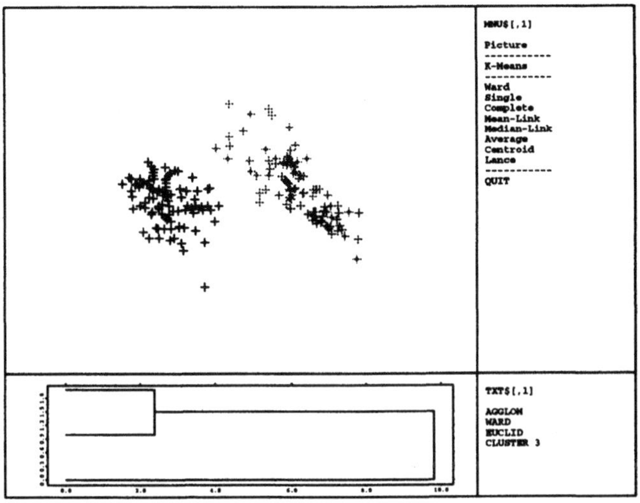

FIGURE 3.3. Principal component plot of Swiss banknote dataset. With the Ward method and the euclidean distance we have chosen 3 clusters. Notice that the clusters do not coincide with the clusters to be found by Eeploratory projection pursuit. To achieve the cluster structure found in exploratory projection pursuit more work has to be done.

with $\#pq$ =number of variables which have the characteristics $x_{i,k} = p$ and $x_{j,k} = q$. Again the choice of the distance depends on the problem. Other distances can be found in Jambu & Lebeaux (1983).

We now have several methods in the agglomerative cluster analysis to merge two clusters. Each method has it own advantages and disadvantages:

Single linkage was developed by Sneath (1957) in the context of taxonomy. The two clusters will be merged when the distance between the two closest neighbours gets minimal. As a result this method tends to produce long chains.

Complete linkage merges two clusters, if the distance between the farthest points in the cluster is minimal. It will produce compact, hyperspherical clusters with highly similar objects.

Average linkage is a compromise of the two methods of Sokal & Michener (1958). Here we take the average of the distances between the farthest and the closest datapoint in the clusters. We will get spherical clusters.

Centroid linkage merges the cluster with nearest distances between the gravity centers of the clusters. This method can be used to find clusters

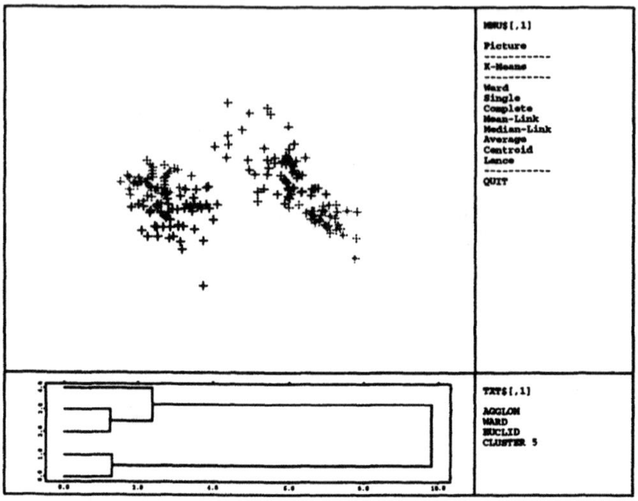

FIGURE 3.4. Principal component plot of Swiss banknote dataset. With the Ward method and the euclidean distance we have chosen 5 clusters.

with different numbers of observations. But large clusters can contain very heterogeneous objects.

Wards method was designed to optimize an objective function, the minimum variance within the clusters (Ward 1963).

Lance- and Williams method is generalization of the preceding methods. By variation of the parameter β we achieve similar results as in the methods before.

Method	α_i	α_j	β	γ
Single linkage	0.5	0.5	0.0	−0.5
Complete linkage	0.5	0.5	0.0	0.5
Average linkage	$\frac{u_i}{u_i+u_j}$	$\frac{u_j}{u_i+u_j}$	0.0	0.0
Centroid linkage	$\frac{u_i}{u_i+u_j}$	$\frac{u_j}{u_i+u_j}$	$\alpha_i\alpha_j$	0.0
Ward	$\frac{u_i+u_l}{u_i+u_j+u_l}$	$\frac{u_j+u_l}{u_i+u_j+u_l}$	$\frac{-u_l}{u_i+u_j+u_l}$	0.0
Lance- and Williams	$\frac{(1-\beta)u_i}{u_i+u_j}$	$\frac{(1-\beta)u_j}{u_i+u_j}$	β	0.0

TABLE 3.1. Constants for computing the new distances for different agglomerative methods. The weights u_i are weights for each cluster, in the simplest case this is the number of observations in each cluster.

Since we need some criteria to decide about the number of clusters we can generate a dendrogram as can be seen in Figure 3.1. The x-axis is the criterion used for merging the clusters. It can be used to decide graphically how many clusters are in the data. No objective approach exists how to choose the number of clusters. One approach was made by Rand (1971) through a measure of correspondence.

3.2 Teachware

Today we have a class of programs that we can call *teachware*. The main aim of teachware programs is to improve the quality of learning statistics for the students. One way to fulfill this aim is to make statistical methods available to the student by

- an individual learning process, where the student determines the speed of learning and
- to learn by playing with statistical techniques.

These aims have a direct influence on the structure of a teachware program. From the second aim it follows that only graphical environments can be used. The program has to be interactive so that the student can get an immediate feedback. Since the student himself determines the speed of learning, he has to be independent from the teacher which means that the program should be very user-friendly. Obviously menu driven environments are necessary as well as good and easily accessible help systems.

We have a lot of techniques in statistics and econometrics which are highly interactive and need graphics (principal component analysis, projection pursuit techniques).

If these aims are fulfilled we can hope to increase the motivation of students to learn and use statistics, especially if they do not have a mathematical background.

Teachers can expect from the teachware an easier and faster understanding of the statistical methods. For example an explanation how the stem and leaf plot is built up will be less informative than the graphical construction shown in CIT. Often it is said that one benefit of teachware is that teachers will save time which they can spend better on methodology. According to my experience this not true as the students have to handle the operating system. We could use teachware programs to test new statistical methods if they were really working well. It might also be a way to make people use new methods. Although the supersmoother is not distributed as a teachware program, the success is connected to the success of `S-Plus` as a statistical programming language.

As a result we have now a variety of teachware programs (Koch & Haag 1995):

CCI 2.7 which concentrates on the concept of confidence intervals and its connections to the sampling distribution, the standard error of an estimate and hypothesis testing,

CIT which shows some basic concepts and explains some graphical techniques (stem and leaf plot, histogram),

First Bayes 1.1 which is designed for teaching and learning elementary Bayesian statistics,

P.C. Convolution which is a tool that helps the students to visualize the convolution and correlation operations,

PRISTAT 1.0 which realizes the statistical procedures and methods of a textbook for introductory applied statistics (Kolev 1993),

SchoolStat 2.0.4 which allows the students to make basic statistics (correlation, linear regression, contingency tables, nonparametric and parametric tests for two samples),

Sila 1.0 which shows students how inferential statistics work and

XploRe 3.2 which has a teachware module where students can see how some nonparametric and some multivariate techniques work (Proenca 1994).

We will shortly describe the aims of teachware module of XploRe 3.2 which was developed by Proenca (1994). The basic assumptions are that

- the learning process should be interactive, exploratory and directed by the student,
- the system should be user-friendly and
- some statistical techniques need graphics and interactivity.

The module handles the visualization of five topics: the reasoning of the ordinary least squares (OLS) rule in linear regression, the impact of influent observations on the OLS rule, principal component analysis, density estimation and nonparametric regression. The program is menu driven and the opening screen is shown in Figure 3.5. We can now choose between the different topics which are all realized in independent macros.

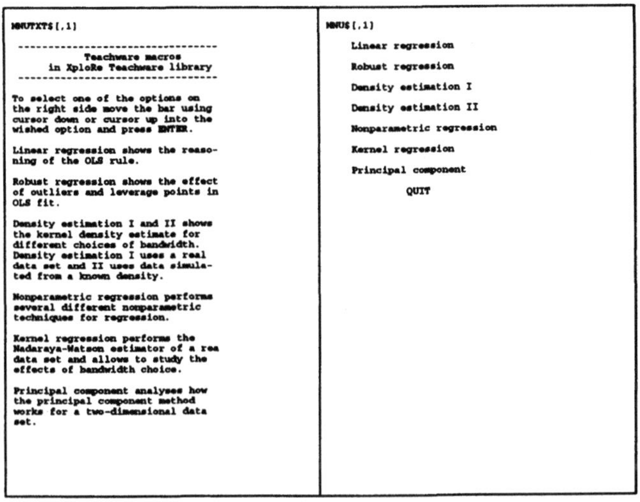

FIGURE 3.5. The opening screen of the teachware library of XploRe 3.2.

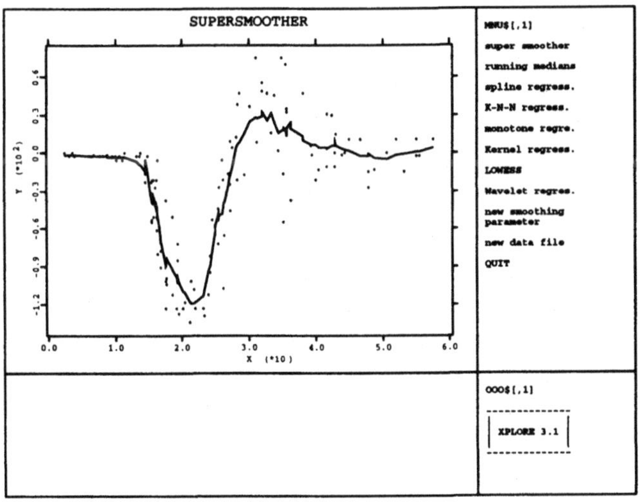

FIGURE 3.6. The screen for nonparametric regression in the teachware library of XploRe 3.2. The supersmoother was chosen for a regression on the motorcycle data set.

Take for example the nonparametric regression and you will get the screen in Figure 3.6. We can choose here between different nonparametric regresssion smoothers:

- the supersmoother,
- the running median,
- the spline smoother
- the k-nearest-neighbour smoother,
- the monotone regression smoother,
- the Nadaraya-Watson smoother,
- the locally weighted regression smoother (LOWESS) and
- the wavelet regression.

The main advantage of this part of the teachware macro is that we can change the smoothing parameter to inspect visually what over- and undersmoothing means. Additionally we can load other datasets to see how the smoother will behave on them. With these possibilities we might be able to demonstrate how different smoother will work and where the advantages and disadvantages of the smoothers are.

But contrary to the basic aims given, the learning process is neither self-explaining nor can it be directed by the student. Without the knowledge how the estimators are defined (there is no help provided on the different estimators) students will be completely confused.

Most of the teachware programs only try to visualize the techniques because a graphic is often much more helpful than an explanation. It is helpful for instructing statistical methods and teachers can use it in lectures.

An approach to build up teachware where the student can really work independently from the teachers is offered by programs like ToolBook for Windows 3.1. The aim of this program is to create a multi-media document, which is composed by graphics, sound and text. A hypertext system makes it possible to build up various levels of difficulty for one topic.

3.3 Regression Methods

3.3.1 Introduction

If we develop a programming language for a statistical software we need to have an idea how differentiated a programming language should be. Obviously we need a lot of procedures in statistics that allow us to manipulate matrices in various ways. But how deeply should we get involved in the statistical routines? We will discuss this for the case of selected (nonparametric) regression methods.

Regression method	Function name		Realization	
	S-Plus	XploRe	S-Plus	XploRe
Alternating conditional expectation	ace	ace	cmd	cmd
Alternating sliced inverse regression		altsir		proc
Additive and variance stabilizing trf.	avas		cmd	
Generalized additive model	gam	dogam	cmd	proc
Generalized linear model	glm	doglm	cmd	proc
	glim		cmd	
Isotonic Regression		isoreg		cmd
k-nearest-neighbour		mknn		cmd
Least median of squares	lmsreg		cmd	
Local polynomial regression		lpregest		proc
		lpregauto		proc
Local regression	loess		cmd	
Locally weighted regression	lowess	lowess	cmd	cmd
Minimum absolute residual	l1fit	l1line	cmd	cmd
Nadaraya-Watson	ksmooth	sker	cmd	cmd
		regauto		proc
		regestp		proc
Projection pursuit regression	ppreg	ppr	cmd	proc
Recursive partition	tree	rpr	cmd	cmd
Robust regression	rreg		cmd	
Running medians	smooth	rmed	cmd	cmd
Smoothing spline	smooth.spline	spline	cmd	cmd
		xspline		proc
Sliced inverse regression		sir		proc
		sirII		proc
Supersmoother	supsmu	supsmo	cmd	cmd
Symmetrized k-nearest-neighbour		sknn		cmd
Wavelet regression	S+Wavelets	wavereg	package	proc

TABLE 3.2. Regression methods in S-Plus and XploRe 3.2.

The two of the most advanced packages for (nonparametric) regression methods are S-Plus and XploRe 3.2 (for an overview see Table 3.2).

The question is now which methods should be implemented as user accessable programs (in XploRe called macros) and which really should be implemented as nonuser accessable commands. We want to state that in S-Plus everything is a command, and except for the wavelets no user has an insight to the details of a regression method. In **XploRe** most of the regression methods are procedures written in the **XploRe** macro language.

The criteria for programming a macro or a command for a regression methods are:

1. Does the regression method follow a fixed algorithm? Do we need any parameters which can not described by a parameter vector?
2. If we choose the form of a macro, is it fast enough? Do we have enough memory to store all necessary data?
3. If we choose the form of a command, can it be built up easily from the basic matrix commands?

3.3.2 Commands versus Procedures

Alternating conditional expectation (ACE). In ACE we try to fit an additive model of the form

$$\theta(y) = \phi_1(x_1) + \ldots + \phi_p(x_p) + \epsilon$$

The measure

$$e^2 = \frac{E[\theta(y) - \sum_{k=1}^{p} \phi_k(x_k)]^2}{E[\phi^2(y)]}$$

of the unexplained variance is used to find optimal solutions for the functions $\hat{\theta}$ and $\hat{\phi}_k$. Breiman & Friedman (1985) who suggested ACE, showed that a nontrivial solution can be constructed by an iterative algorithm (here for the univariate case):

$$\begin{aligned}
\phi^{(1)}(x) &= E[y|x] \\
\theta^{(1)}(y) &= E[\phi^{(1)}(x)|x] \\
\phi^{(2)}(x) &= E[\theta^{(2)}(y)|x] \\
\theta^{(2)}(y) &= E[\phi^{(2)}(x)|x] \\
&\ldots
\end{aligned}$$

We need good estimators for the functions $\phi^{(j)}$ and $\theta^{(j)}$. In **S-Plus** the supersmoother is used, but we can use any univariate regression method. So the ACE routine should be a macro.

Additive and variance stabilizing transformation (AVAS). The AVAS method can be seen as a modification of ACE. It tries to stabilize the variance

$$Var(\theta(y)|\sum_{i=1}^{k}\phi_i(x_i)) = const.$$

Again we can use several univariate regression smoothers. We would like to have this regression method as a macro.

Average derivative estimation (ADE). The aim of ADE is to fit the average slope of the unknown regression function:

$$\delta = E(\nabla m(x))$$

with $\nabla m(x)$ the partial derivatives of the unknown regression function $m(x)$, especially if we assume that the unknown regression function can be written as

$$m(x) = g(x^T\alpha)$$

with g an unknown univariate function and α is a projection vector. It follows that

$$\nabla m(x) = \nabla g(x^T\alpha)^T\alpha$$

which under suitable conditions leads to

$$\delta = E\left(-y\frac{\nabla f(x)}{f(x)}\right).$$

The problem that the density $f(x)$ becomes small at the borders can be solved via a weighted average derivative estimation (WADE). For estimating the unknown multivariate density and its derivatives several methods are possible, e.g. kernel density estimation or density estimation via orthogonal polynomials.

An implementation can be found in Kötter & Turlach (1995). The implementation is easy and quick, and other density estimators can be used in principle. An implementation as a procedure is necessary.

Generalized additive models, generalized linear models (GAM, GLIM). The model used in GLIM is

$$\theta(y) = x^T \alpha$$

with α a projection vector and θ a link function that has to be specified. The number of the link functions and the number of appropriate algorithms has been increased over the time. Although this model is completely parametric, the extensibility requires programming as a procedure.

The model fitted in GAM is

$$\theta(y) = \alpha + \sum_{k=1}^{p} \phi_k(x_k).$$

Again the functions $\theta(y)$, ϕ_k can be any univariate regression function. In Kötter & Turlach (1995) in Figure 11.2 -11.4 different regression methods (supersmoother, local linear and s-k-nearest-neighbour) are used on the same dataset. Again programming as a procedure is required.

Isotonic regression. The isotonic regression on a set $\{X_i, Y_i\}$ is found by least squares minimization

$$\frac{1}{n}\sum_{i=1}^{n}(Y_i - \hat{\phi}(X_i))^2$$

subject to the constraint $X_i \leq X_j \rightarrow \hat{\phi}(X_i) \leq \hat{\phi}(X_j)$. The algorithm is described in Härdle (1990) and fixed. An implementation as a command is prefered.

***k*-nearest-neighbour regression, running median**. The k-nearest-neighbour regression estimator can be defined as

$$\hat{\phi}(x) = \frac{1}{n}\sum_{i=1}^{n} W_{ki}(x) Y_i$$

with weights $W_{ki} = n/k$, if $i \in J_x$ and $W_{ki} = 0$ otherwise. J_x is defined as the set $\{: X_i$ is one of the k nearest observations to $x\}$.

A variation in the univariate case is the symmetric k-nearest-neighbour estimator. For each datapoint X_i we have to take the same amount of datapoints on the left and right side of the datapoint.

The running median takes the median of the observations Y_{J_x}.

All algorithms for estimation are fixed and we can implement them as commands.

Least median of squares, minimum absolute residual (LMS, MAD). The LMS is a robust version of a least squares fit (Rousseeuw 1994). Instead of minimizing the average of the squared residuals

$$\frac{1}{n}\sum_{i=1}^{n}(Y_i - \hat{\beta}_0 - X_i^T\hat{\beta})^2$$

in the model

$$Y_i = \beta_0 + X_1^T\beta$$

we minimize

$$\text{median}\{(Y_1 - \hat{\beta}_0 - X_1^T\hat{\beta})^2, ..., (Y_n - \hat{\beta}_0 - X_n^T\hat{\beta})^2\}$$

by optimizing $\hat{\beta}_0$ and $\hat{\beta}$. In MAD we optimize the model by the average absolute residuals

$$\frac{1}{n}\sum_{i=1}^{n} | Y_i - \hat{\beta}_0 - X_i^T\hat{\beta} | .$$

The smoothing procedures should be implemented as commands as some optimization has to be done.

Local polynomial regression. One of the drawbacks of the Nadaraya-Watson estimator (see below) is that the fit becomes less accurate at the boundary. This is due to the fact that we have fewer observation to average at the boundaries than in the interior of our data. One solution is the use of "boundary kernels".

But it is reasonable to use another class of estimators, the local polynomial estimator. Following Fan, Gasser, Gijbels, Brockmann & Engel (1993) it can be written as the minimization problem

$$\sum_{i=1}^{n}\left(Y_i - \sum_{j=0}^{p} b_j(x_0)(X_i - x_0)^j\right)^2 K\left(\frac{X_i - x_0}{h}\right)$$

with K a kernel function and h a smoothing parameter, the bandwidth. Here we approximate the unknown regression function m locally by a polynomial of order p. The Nadaraya-Watson estimator can be regarded as a local constant fit. For a detailed description of the theory and the methodology of the local polynomial regression see Fan & Gijbels (1996).

The more important aspect for a statistical software system is to implement it in a efficient way. If we define

$$
\begin{aligned}
X &= \begin{pmatrix} 1 & (X_1 - x_0) & \dots & (X_1 - x_0)^p \\ \vdots & \vdots & \dots & \vdots \\ 1 & (X_n - x_0) & \dots & (X_n - x_0)^p \end{pmatrix} \\
Y &= \begin{pmatrix} Y_1 \\ \vdots \\ Y_n \end{pmatrix} \\
\hat{b}(x_0) &= \begin{pmatrix} \hat{b_0}(x_0) \\ \vdots \\ \hat{b_p}(x_0) \end{pmatrix} \\
W &= diag\left(K\left(\frac{X_i - x_0}{h}\right)\right)
\end{aligned}
$$

we can calculate the coefficients $b_j(x_0)$ by

$$
\begin{aligned}
\hat{b}(x_0) &= (X^T W X)^{-1} X^T W Y \\
&= S(x_0)^{-1} T(x_0) \\
&= \begin{pmatrix} S_0(x_0) & S_1(x_0) & \dots & S_p(x_0) \\ \vdots & \vdots & & \vdots \\ S_p(x_0) & S_1(x_0) & \dots & S_{2p}(x_0) \end{pmatrix}^{-1} \begin{pmatrix} T_0(x_0) \\ \vdots \\ T_p(x_0) \end{pmatrix} \\
S_j(x_0) &= \sum_{i=1}^{n} K\left(\frac{X_i - x_0}{h}\right)(X_i - x_0)^j \\
T_j(x_0) &= \sum_{i=1}^{n} K\left(\frac{X_i - x_0}{h}\right)(X_i - x_0)^j Y_i.
\end{aligned}
$$

We can implement it quickly by

1. Compute for all datapoints X_i the functions $S_j(X_i)$ and $T_j(X_i)$
2. Compute for all datapoints X_i the inverse of $S(X_i)$ and multiply it with $T(X_i)$
3. Since we approximate in each datapoint by a polynomial we only need $\hat{m}_h(X_i) = b_0(X_i)$

An estimation of the l-th derivative can be computed by the use of $b_l(X_i)$. Seifert, Brockmann, Engel & Gasser (1994) have developed a fast and general algorithm which works in $O(n)$ operations.

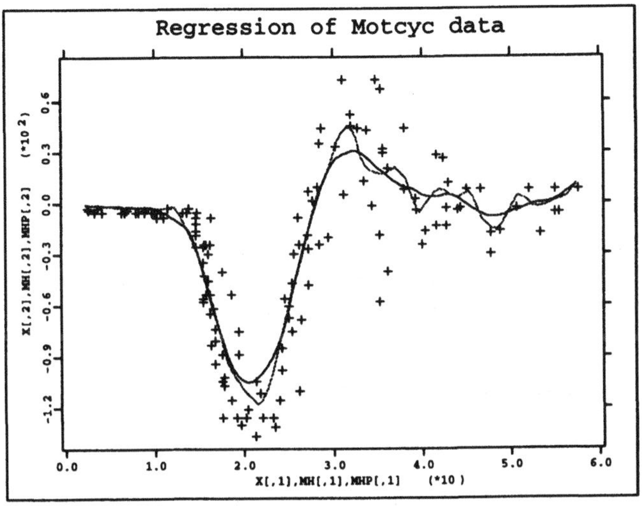

FIGURE 3.7. Regression on the motorcycle data with Nadaraya-Watson estimator (gray) and local linear polynomial estimation (black). Both bandwidths are crossvalidated $h_{NW} = 2.4$, $h_P = 5$.

The local polynomial regression should be implemented as a procedure since the kernel is not fixed. But the algorithm of Seifert et al. (1994) can only be implemented as a command since it is very complicated.

Local regression, locally weighted regression and supersmoother, robust regression (LOESS, LOWESS, SUPSMO, M-estimator). All these methods are very specialized algorithms. It is preferred to implement them as commands rather than as a macro.

Nadaraya-Watson estimator. The Nadaraya-Watson estimator can be seen as a generalization of the regressogram.

$$\hat{m}_h(x) = \frac{\sum_{i=1}^{n} K_h(x - X_i) Y_i}{\sum_{i=1}^{n} K_h(x - X_i)}$$

with (X_i, Y_i) being data sampled from the model

$$y = m(x) + error.$$

The function $K_h(x) = K(x/h)/h$ is a kernel function and m the unknown regression function. The amount of smoothing is given by the bandwidth h. In Table 4.3 some kernel functions are given. For the multivariate case product kernels are also used

$$K(x_1, x_2, ..., x_d) = K(x_1)K(x_2)...K(x_d)$$

with support $[-1, 1]^d$ instead of the unit ball as given in Table 4.3.

The Nadaraya-Watson, and kernel estimators in general, are very popular under the (nonparametric) smoothing methods since they can easily be understood intuitively. They are easy to implement. Binning or the fast fourier transformation allows efficient implementations.

To derive the mathematical properties for the Nadaraya-Watson estimator is a little bit tricky. Detailed explanations and references can be found in Härdle (1990).

As already seen in the part of exploratory projection pursuit, the choice of the smoothing parameter is a crucial task. Here an interactive environment can help.

In Figure 3.8 the Nadaraya-Watson estimator is computed for the trivariate case. Here interactive contouring is required to get the shape of the function

$$\hat{m}_h(x) = const.$$

The implementation as a procedure is necessary as we have the choice of different kernels and as we can use binning methods. For example `S-Plus` supports only the univariate Nadaraya-Watson estimator and offers only 4 different kernels.

Projection pursuit regression (PPR). The general problem of estimating a response function

$$E(Y|X = x) = m(x_1, ..., x_p)$$

with (X, Y) being a pair of random variables with $X \in I\!R^p$ and $Y \in I\!R$ is the "curse of dimensionality". A direct approach with (nonparametric) methods will lead to a strong oversmoothing since the sparseness of the space will require to include a lot of space to do a local averaging. To estimate the response function $\hat{m}_M(x)$ from the data $\{(x_1, y_1), ..., (x_n, y_n)\}$ the following idea seems to be much more attractive:

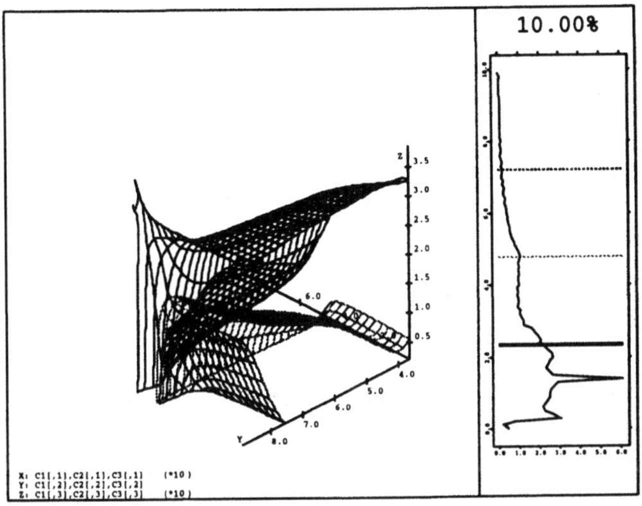

FIGURE 3.8. Trivariate regression with the Nadaraya-Watson estimator. The variable MEDV is regressed on the variables CRIM (crime rate), RM (mean rooms per house) and LSTAT (percentage of people in lower status). In each variable the binwidth was chosen so that we have 25 bins, the bandwidth was chosen as 12× the binwidth and the quartic product kernel was used. The lower (dark gray) surface represents a median price of approximately 30.000 US$, the middle (light gray) surface approximately 21.000 US$ and the upper (black) surface approximately 13.000 US$.

1. Set $r_i^{(0)} = y_i$.

2. Minimize the error function $E_j = 1 - \frac{\sum_{i=1}^{n} (r_i^{(j-1)} - \hat{m}_j(\hat{\alpha}_j^T x_i))^2}{\sum_{i=1}^{n} (r_i^{(j-1)})^2}$ by varying over the parameter $\hat{\alpha}_j \in I\!R^p$ and $\hat{m}_j$ a univariate smooth function.

3. Define $r_i^{(j)} = r_i^{(j-1)} - \hat{m}_j(\hat{\alpha}_j^T x_i)$ and repeat step 2 until E_j becomes small.

This algorithm leads to an estimation of the response function by

$$\hat{m}_M(x) = \sum_{j=1}^{M} \hat{m}_j(\hat{\alpha}_j^T x).$$

The advantages of estimating the response function in this way are:

1. We can use univariate regression functions instead of their multivariate analogues and avoid the "curse of dimensionality".
2. Univariate regression are easy and quick to calculate.
3. In contrast to generalized additive models (GAM) we are able to model interaction terms.
4. Unlike local averaging methods, e.g. k-nearest-neighbour methods, we are able to ignore information poor variables.

Of course we have some disadvantages with this model too:

1. We have to examine a p dimensional parameter space to estimate $\hat{\alpha}_j$.
2. We have to solve the problem of selecting a smoothing parameter if we use nonparametric smoothers for $\hat{m}_j$.
3. The interpretation of a single term is not as easy as in GAM.

With the assumption of standardized predictor and response variables, that is

$$E(X_i) = 0, Var(X_i) = 1, E(Y) = 0, Var(Y) = 1$$

and the error function E_j Friedman & Stuetzle (1981b) constructed a special smoother for the unknown regression function $\hat{m}_j$. The method is very similar to the well known supersmoother. The smoothing algorithm is the following (h a smoothing parameter):

FOR each X_i DO

$Z_x = \{X_j \mid X_i - h \leq X_j \leq X_i + h\}$
$Z_y = \{Y_j \mid X_i - h \leq X_j \leq X_i + h\}$
$\beta = Linear - Regression(Z_x, Z_y)$
Do a local linear regression to reduce the bias.
$\sigma_i = 1/\#Z_x \sum_{Z_x} (Z_y - \beta_1 Z_x - \beta_0)^2$
$\Sigma_i = \{\sigma_i \mid X_i - h \leq X_j \leq X_i + h\}$
$h_i = mean(\Sigma_i)$
Choose a variable bandwidth to avoid spurious fits along existing directions

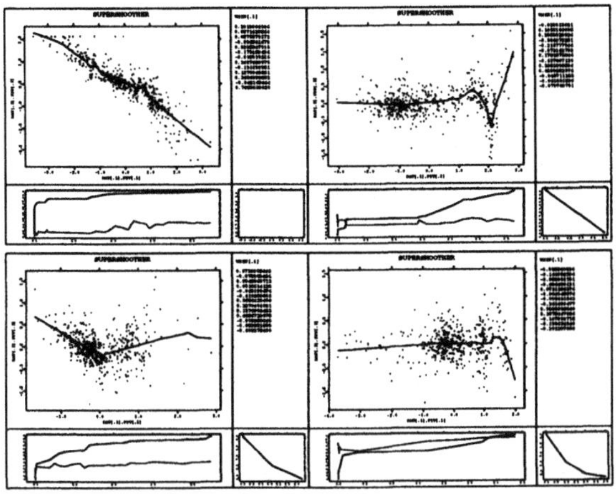

FIGURE 3.9. Four-term projection pursuit regression with the Boston housing data. The regression function used is the supersmoother.

ENDO

FOR each X_i DO

$W_i = median(Y_{i-1}, Y_i, Y_{i+1})$

Do the running median to protect against isolated outliers.

$W_x = \{X_j \mid X_i - h_i \leq X_j \leq X_i + h_i\}$ without $\{X_i\}$

$W_y = \{Y_j \mid X_i - h_i \leq X_j \leq X_i + h_i\}$ without $\{Y_i\}$

$\beta = Linear - Regression(W_x, W_y)$

$\hat{Y}_i = \beta_1 X_i + \beta_0$

ENDO

Moreover Friedman & Stuetzle (1981b) suggest to use backfitting to improve the quality of the estimate. Backfitting means here that a new fit will be done along the existing directions if we have determined a new direction.

Again other univariate regression functions can be used. A theoretical result with PPR and the Nadaraya-Watson estimator is obtained by Hall (1989b).

Since each univariate regression function can be used, an implementation as a procedure is required. The implementation in `XploRe` as a macro which will be interpreted, needs five minutes for each term of a fit, whereas in `S-Plus` the same fit including the backfitting only needs seconds as the command is written in Fortran.

Smoothing splines. The smoothing spline tries to minimize

$$\frac{1}{n}\sum_{i=1}^{n}(Y_i - \phi(X_i))^2 + \lambda \int_a^b \phi^{(m)}(x)dx$$

with a smoothing parameter λ. The solution to this problem is a (smoothing) spline (piecewise polynomials) of order $m+1$. This is also a very specialized algorithm and an implementation as a command is required.

Regression partition trees (RPR). The idea is to estimate

$$\hat{\phi}(x) = \sum_{j=1}^{q} c_j I(x \in N_j)$$

so that the N_j are disjoint hyperrectangles which cover the whole $\mathbb{R}^p$. The coefficients can be estimated by

$$\hat{c}_j = \frac{1}{n_j} \sum_{X_i \in N_j} Y_j$$

with n_j the number of observations which fall in N_j. The N_j are obtained by splitting along the coordinate axes such that the residual sum of squares is minimized with such a split. The chosen splittings can be visualized by a tree. For further details see Breiman et al. (1984).

This algorithm also is very specialized and should be implemented as a command.

Sliced inverse regression (SIR). SIR is a method proposed by Duan & Li (1991). It is based on the model

$$y = \phi(\alpha_1^T X, ..., \alpha_d^T X)$$

with an unknown function regression ϕ, unknown parameter d and $\alpha_1, ..., \alpha_d$ unknown projection vectors. The idea is to use the inverse regressions function ϕ^{-1} which consists of d univariate regression. Li (1991) showed that under certain conditions the inverse regression function is lying in a linear subspace of $\alpha_1, ..., \alpha_d$. It holds that the covariance of the inverse regression functions is degenerated in each direction orthogonal to $\alpha_1, ..., \alpha_d$. So we can construct the following algorithm:

1. Divide the range of y into slices.
2. Compute the inverse regression function, e.g. take the mean of X_i.
3. Compute the covariance of the inverse regression function

4. Decompose the covariance matrix by eigenvectors and eigenvalues
5. Throw away the eigenvectors with the smallest eigenvalue to obtain d

We can identify the important subspace for the regression and get a dimension reduction. Now a multivariate regression method can be used to find the unknown regression function. A more complicated method, called SIRII, is suggested by Li (1992) which is looking at the conditional covariance $Cov(X \mid y)$. Another method for a multivariate y is proposed by Li, Aragon & Thomas-Agnan (1995) which uses an alternating sliced inverse regression.

The algorithms above are easy to program and work very fast, so an implementation as a procedure is recommended.

Wavelet regression. The wavelet estimation relies on a special orthonormal function system. In the Fourier system the basis functions are periodic and can be located only once. We can only use the different frequencies for the estimation from the orthonormal function system. The wavelet basis functions have to be fixed in location and frequency. An estimate of an univariate regression function is

$$\hat{m}(x) = \sum_{k=0}^{2^{j_0-1}} \hat{a}_k \varphi_{j_0,k}(x) + \sum_{j=j_0}^{j_1} \sum_{k=1}^{2^j} \hat{b}_{j,k} \psi_{j,k}(x)$$

where it holds $\varphi_{j_0,k}(x) = 2^{j_0/2}\varphi(2^{j_0}x - k)$, $\psi_{j,k}(x) = 2^{j/2}\psi(2^j x - k)$ and $\varphi(x)$ the father wavelet and $\psi(x)$ the mother wavelet. This pair of functions is sufficient to characterize a whole set of functions which forms a orthonormal basis of $L_2(\mathbb{R})$. Different basis systems with compact support are possible:

1. Haar basis, Daubechies-k
   compactly supported and highest number of vanishing moments for the mother wavelets compatible with their support width
2. Symmlet-k
   compactly supported and least asymmetric
3. Coiflet-k
   compactly supported and a high number of vanishing moments for the father and mother wavelets

From the orthogonality we can estimate the constants from the data (X_i, Y_i)

$$\hat{a}_{p,q} = \int_{\mathbb{R}} \hat{m}(x)\varphi_{p,q}(x)dx$$

$$= \frac{1}{n}\sum_{i=1}^{n} Y_i \varphi_{p,q}(X_i)$$

and

$$\begin{aligned}\hat{b}_{p,q} &= \int_{I\!R} \hat{m}(x)\psi_{p,q}(x)dx \\ &= \frac{1}{n}\sum_{i=1}^{n} Y_i \psi_{p,q}(X_i)\end{aligned}$$

To estimate the integrals accurately we need to have a equidistant design of the X_i. For further details see Härdle, Kerkyachrian, Picard & Tsybakov (1995).

The estimation itself will exactly reproduce the Y_i's if we use all coefficients, and the resulting curve will not be smooth. Up to now the estimation is an interpolation algorithm. Since we expect that the true regression curve $m(x)$ is overlayed with noise, the idea is to cut the smallest coefficients $b_{j,k}$. The standard techniques for this are called hard and soft thresholding. In the hard thresholding with a threshold t we replace the coefficients by

$$\hat{b}^H_{j,k} = \hat{b}_{j,k} I(|\, \hat{b}_{j,k} \,| > t)$$

and in the soft thresholding by

$$\hat{b}^H_{j,k} = \text{sign}(\hat{b}_{j,k}) \mid \hat{b}_{j,k} - t \mid I(|\, \hat{b}_{j,k} \,| > t).$$

The choice of t is the same crucial problem as the bandwidth selection in the kernel regression.

From Donoho, Johnstone, Kerkyacharian & Picard (1995) and Neumann (1994) we have another suggestion for the regression case. They propose an individual threshold for each coefficient:

$$t_{j,k} = \sqrt{2 s_{j,k} \log(\#\text{thresholded coefficients})}$$

with $s_{j,k}$ the variance of the coefficient $b_{j,k}$.

The Figures 3.10 - 3.13 show a wavelet regression for the variables FA and FP with the Daubechies-4 wavelet as basis. Since equidistance is needed, we have used a 1-nearest-neighbour smooth to fill up the gaps. This leads to the horizontal lines in the right of Figure 3.11. Figure 3.12 and 3.13 show

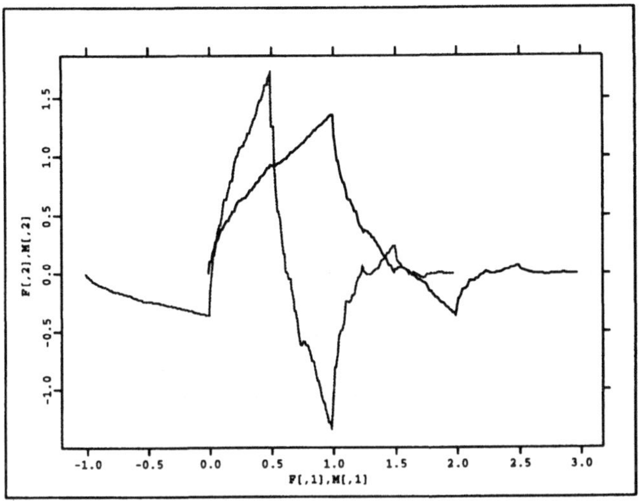

FIGURE 3.10. Father wavelet (left) and mother wavelet (right) of Daubechies-4.

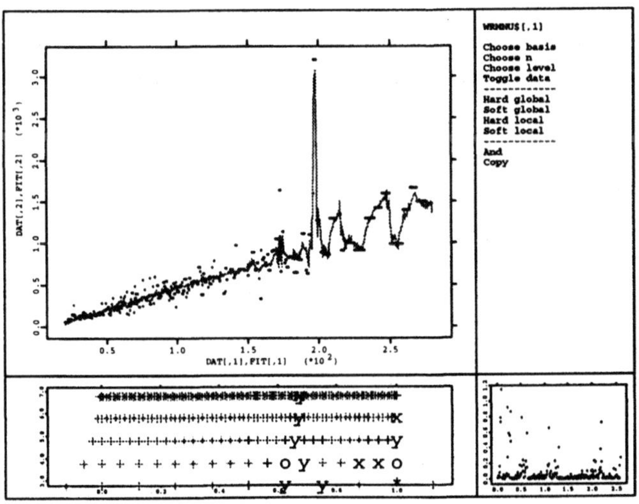

FIGURE 3.11. Wavelet regression of FA against FP with Daubechies-4. Global hard threshold ($t > 8$).

the original dataset whereas Figure 3.11 shows the smoothed dataset which is used for the estimation. We see that in Figure 3.11 and 3.12 neither the global nor the local hard threshold is able to exclude the outlier. The plot

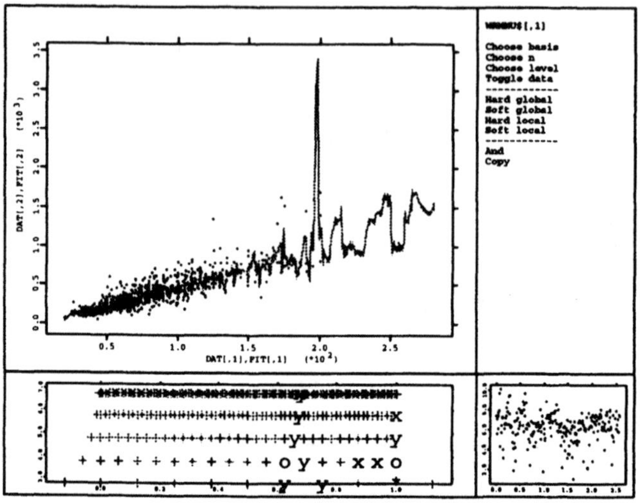

FIGURE 3.12. Wavelet regression of FA against FP with Daubechies-4. Local hard threshold ($t > 7$).

beyond the fit shows us the wavelet coefficients; red the ones included in the estimation, blue the ones not included. Finally we brushed the coefficients interactively to kill the peek, and this results in Figure 3.13 which shows a much nicer fit.

A fast implementation of the computation of the wavelet regression can be done via the "cascade"' algorithm:

$$
\begin{aligned}
b_{j,k} &= \sum_l (-1)^{1+l-2k} h_{1+2k-l} a_{j+1,l} \\
a_{j,k} &= \sum_l h_{l-2k} a_{j+1,l}.
\end{aligned}
$$

The coefficients h_l are depended on the wavelet basis chosen and can be found in Daubechie (1992). In Daalhuis (1992) the last relation and the computation of the mother and father wavelet is described via an iterative algorithm:

$$
\begin{aligned}
\varphi_1(x) &= \begin{matrix} 1 & \text{if } |x| < 0.5 \\ 0 & \text{otherwise} \end{matrix} \\
\varphi_r(x) &= \sqrt{2} \sum_{k=-\infty}^{\infty} h_k \varphi_{r-1}(2x - k)
\end{aligned}
$$

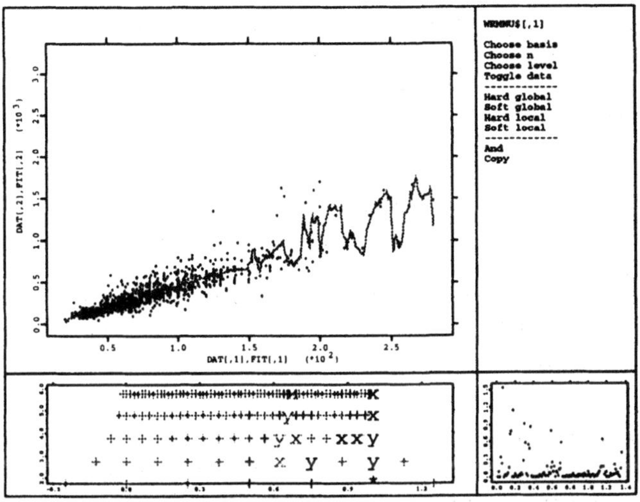

FIGURE 3.13. Wavelet regression of FA against FP with Daubechies-4. Global hard threshold ($t > 10$) and the waves caused by observation 1323 killed interactively.

$$\psi_r(x) = \sqrt{2} \sum_{k=-\infty}^{\infty} (-1)^k h_{1-k} \varphi_r(2x-k)$$

As a consequence the algorithm can be decomposed into three parts:

- Estimation of the coefficients
- Thresholding
- Estimation of the regression curve

which can be done in procedures.

Conclusion. As we have seen most of the univariate (nonparametric) methods can be implemented as commands. Nevertheless if we want to choose a smoothing parameter we will automatically run into trouble.

The problem is that we want to avoid to make black-box-algorithms. Two examples:

1. The spline algorithm in **XploRe**. We implemented the method of Silverman (1984) for automatic smoothing. Prof. Thomas-Agnan recognized that the automatic parameter choice does not work as described in

Silverman. Since the code was translated from Fortran to Pascal for XploRe 2.0 and later from Pascal to C for XploRe 3.0 we made somewhere a misstransfer. This was the reason to implement the smoothing spline as a macro in XploRe 3.2 (Buys 1995).
If we make an error in the implementation a specialist will be able to correct this if the code is accessible.

2. The PPR command in S-Plus. First we do not know how the algorithm really works internally. We only have the help for the PPR command where the user is referred to the original paper of Stuetzle and Friedman. We think that it would be a far better advise to direct the user to the paper of Friedman (1985) since the version in S-Plus supports multi-response-models.

If we have a simple univariate smoothing method we can program it as a command, otherwise we need a macro. Nearly all multivariate regression methods need univariate regression methods, so they should be implemented as macros.

Obviously we have to judge how fast a method works and how much storage will be needed. If it takes too much time or needs too much memory then we have to implement the method as a command. But we know that in the future the computers will be much faster than today.

4

Exploratory Projection Pursuit

Summary

In this chapter we will discuss in detail one statistical technique. We will cover possible extensions (multivariate projections, inclusion of discrete variables) and show which graphics are used for representing results. The danger in EPP is that we interpret a random structure as a real structure in the data. We describe tests for detecting a structure which is not a multivariate gaussian distribution. Pictures are presented from tools (macros in `XploRe` and `XGobi`) which are used to execute exploratory projection pursuit. At the end we will specify the requirements necessary for a tool to do EPP, but neiter my implementation nor `XGobi` can satisfy all of them.

4.1 Motivation and History

4.1.1 Introduction

The analysis of multivariate data is a problem in statistics being both interesting and difficult. The difficulties have several reasons:

- structures which have a dimension larger than three cannot easily be visualized for human perception
- parametric models which fix the structure can only tell us if the structure is in the data or not, but nothing else
- nonparametric models which do not fix the structure in the data suffer from the "curse of dimensionality", which means we need too many observations to be sure that the estimate is correct

How can we examine structure (here: the distribution) of the data and describe them? The answer is given to us by the theorem of Cramér-Wold (Mardia, Kent & Bibby 1979):

The distribution of a random p-vector X is completely determined by the set of all one dimensional distributions of linear combinations $\alpha^T X$, where $\alpha \in \mathbb{R}^p$ ranges through all fixed p-vectors. (4.1)

That implies that a multivariate distribution can be defined completely by specifying the distribution of all its projections. Of course if we fix the distribution in all two dimensional projections of a multivariate distribution, we determine the multivariate distribution too. Thus we have to look at all two dimensional projections, as it is done in the grand tour (Asimov 1985). But when the dimensions grow we need a long time to review a dense set of projections.

In the next sections we describe some of the techniques which are related to EPP or selected direct predecessors.

4.1.2 Principal Component Analysis

Definition

Principal component analysis tries to summarize multivariate data by principal components. The method was originated by Pearson (1901). The idea is that the first principal component explains as much as possible of the total sample variance. The second principal component explains as much as possible from the unexplained total sample variance. This process can be iterated so that we get p principal components for a sample of a p dimensional random variable X which would explain the total sample variance. Of course we would like to reduce the dimension of the sample to get only the important principal components.

A detailed mathematical description of principal component analysis can be found in books for multivariate statistical analysis, e.g. in Mardia et al. (1979), Morrison (1976) or Härdle & Simar (1995). Since the principal component analysis is a decomposition of the covariance matrix

$$
\begin{aligned}
\Sigma &= \Gamma^T diag(\lambda)\Gamma \\
\Sigma^{-1/2} &= \Gamma^T diag(\lambda)^{-1/2}\Gamma \\
Y &= \Sigma^{-1/2}(X-\mu)
\end{aligned}
$$

with the eigenvectors Γ of Σ which are the principal components and the eigenvalues λ. Y are the data transformed on the principal component axes. The variance explained by the i-th eigenvector given in percent is

$$\frac{\lambda_i}{\sum_{i=1}^{p} \lambda_i}.$$

In fact we are building equivalence classes of (sample) distributions by the covariance matrix. A representant of such an equivalent class would be a gaussian distribution with a covariance matrix Σ. Thus to understand how principal components work it will be enough to study gaussian distributions.

How many components to choose?

The main problem in principal component analysis is to choose how many components should be included so that enough variance is explained. In Mardia et al. (1979) we find several criteria for this:

- **Elbow criterion**
  The elbow criteria are visual criteria. We are looking in a scree plot (Figure 4.1) for an elbow. The scree plot contains the eigenvalues plotted according to size. Since the covariance matrix is semi-positive definite and symmetric we have only nonnegative eigenvalues.

  In Figure 4.1 the first three eigenvalues build up an elbow. For this reason we would include the first two components.

- **90 percent criterion**
  Include so many components that 90% of the total sample variance is explained.

- **Kaiser criterion**
  Exclude those components whose eigenvalues are below the average.

Testing components

The advantage of the PCA as a dimension reduction technique over the factor analysis is that we can make tests and compute confidence intervals for the eigenvalues.

We can calculate a $(1-\alpha)\%$ asymptotic confidence interval by

$$\frac{\hat{\lambda}_i}{1 + z_{0.5\alpha}\sqrt{2/(n-1)r}} \leq \lambda_i \leq \frac{\hat{\lambda}_i}{1 - z_{0.5\alpha}\sqrt{2/(n-1)r}}$$

with λ_i the true eigenvalue, $\hat{\lambda}_i$ the eigenvalue computed by the sample, r the

```
FACTOR SCREE PLOT

EIGENVALUES

   -+----------+----------+----------+----------+----------+-
 4 +                                                       +
   I                                                       I
   I                                                       I
   IA                                                      I
 3 +                                                       +
   I                                                       I
   I                                                       I
   I                                                       I
 2 +                                                       +
   I                                                       I
   I                                                       I
   I                                                       I
 1 +          A                                            +
   I                                                       I
   I                                                       I
   I                     A          A          A          AI
 0 +                                                       +
   -+----------+----------+----------+----------+----------+-
    1          2          3          4          5          6
                              INDEX
```

FIGURE 4.1. Scree plot for Swiss banknote dataset from SYSTAT. We see immediately that one or two components are enough to explain the variance of the dataset.

multiplicity of the eigenvalue, n the number of observations and z_α the $\alpha\%$ quantile of the standard normal distribution.

Figure 4.1 suggests that the last four eigenvalues are almost similar. Anderson (1963) has examined the hypothesis:

$$H_0 : \lambda_{q+1} = ... = \lambda_{q+r}.$$

A likelihood-ratio criterion leads to the statistic

$$-(n-1) \sum_{i=q+1}^{q+r} \log \hat{\lambda}_i + (n-1) r \log \left(\frac{1}{r} \sum_{i=q+1}^{q+r} \hat{\lambda}_i \right)$$

which is asymptotically χ^2 distributed with $r(r+1)/2-1$ degrees of freedom.

A tool for principal component analysis

A tool which does the principal component analysis should offer the following possibilities:

- a scree plot of the eigenvalues including confidence bands
- a plot of the first principal components (as far as possible)
- the sample variance explained by the components
- the possibility to make the tests described above
- to show the eigenvectors in a way that we can see which variables have the greatest influence upon the eigenvector and help us to interpret the components.

4.1.3 Grand Tour

The grand tour was developed by Asimov (1985). The representation of multivariate data is done by showing a sequence of bivariate projections of these data. The theorem of Cramér-Wold is the basis for this method. Asimov proposed the following important properties for the sequence of projections:

- The sequence of projections should become dense in the space of all projections. $G_{2,p}$ ("Grassmannian manifold") stands for the space of all unoriented planes in the p dimensional space.
- The sequence of projections should become dense rapidly in $G_{2,p}$.
- The sequence of projections should become dense uniformly in $G_{2,p}$.
- The sequence of projections should be continuous, which means that the planes before and after the actual projection should be close.
- The sequence of projections should incorporate a degree of flexibility to optimize the goals mentioned above.
- The sequence of projections should be to reconstruct easily .

To fulfill these goals Asimov described three methods to choose a path through $G_{2,p}$.

Torus method
A curve

$$\begin{array}{rcl} \alpha: \mathbb{R} & \rightarrow & T^m \\ t & \rightarrow & (\alpha_1(t), ..., \alpha_m(t)) \\ t & \rightarrow & (\lambda_1 t \equiv 2\pi, ..., \lambda_m t \equiv 2\pi) \end{array}$$

has in the m dimensional torus T^m a dense image if the coefficients $\lambda_1, ..., \lambda_m$ are linearly independent over integers. The m-vector produced by $\alpha(t)$ can be used to construct a matrix which describes the rotation of the p dimensional coordinate system ($m = p(p-1)/2$). Thus each entry describes a rotation between two axes by

$$R_k(t) = \begin{pmatrix} 1 & \dots & 0 & \dots & 0 & \dots & 0 \\ \vdots & \ddots & \vdots & & \vdots & & \vdots \\ 0 & \dots & \cos(\alpha_k(t)) & \dots & \sin(\alpha_k(t)) & \dots & 0 \\ \vdots & & \vdots & \ddots & \vdots & & \vdots \\ 0 & \dots & -\sin(\alpha_k(t)) & \dots & \cos(\alpha_k(t)) & \dots & 0 \\ \vdots & & \vdots & & \vdots & \ddots & \vdots \\ 0 & \dots & 0 & \dots & 0 & \dots & 1 \end{pmatrix}$$

The final rotation matrix will be composed by

$$R(t) = R_1(t)...R_{p(p-1)/2}(t)$$

with $R_1(t)$ the rotation between axis 1 and 2, $R_2(t)$ the rotation between axis 1 and 3, ... and $R_{p(p-1)/2}(t)$ the rotation between axis $p-1$ and p. The projections planes are produced by $(R(t)e_1, R(t)e_2)$. With the appropriate choices of λ_k (e.g. $\lambda_k = \sqrt{k\text{-th prime number}}$ or $\lambda_k = \exp(k) \equiv 1$) the sequence of projections becomes dense in $G_{2,p}$. But the sequence of projections is not uniform distributed even if t is chosen as $l \times$ *step*. The *step* allows us to optimize between the goals of continuity and rapidity. If *step* is large we receive rapidity and if *step* is small we achieve continuity. Of course every projection is reconstructable.

At-random method

Generate two random vectors and orthonormalize them with the Gram-Schmidt method. This can be done in such a way that the sequence of projections is uniformly distributed in $G_{2,p}$. The sequence of projection will become dense uniformly and rapidly, but will not be continuous. Since we are using random seeds to initialize the random generator we can also reconstruct a single projection.

Buja, Asimov & Hurley (1989) have shown a way to obtain continuity. They constructed a subspace interpolation between two projections in

p-space. The idea is to construct a rotation between the two planes in a four dimensional space, given by the projection planes.

This is the preferred method for implementing the grand tour in statistical software.

At-random walk

Asimov before proposed a mixture of these two methods as follows:

- Choose a measure μ on all rotations of the p dimensional coordinate system such that it generates a dense subset in the space of rotations.
- Start with $R(0) = I_p$.
- Generate a rotation g_t according to the law μ.
- Compute $R(t)$ as $g_t R(t-1)$ and generate the projection plane as $(R(t)e_1, R(t)e_2)$.

He described two measures that fulfill these conditions.

The problem of the grand tour is that we will have to review many planes for to find any structures. In Huber (1985) the RANDU dataset is used to show that a rotation by five degrees will hide the structure. The RANDU dataset was generated by a random generator proposed by IBM in the early seventies. As any linear congruential random number generator, it has the property that the data are lying on hyperplanes if we produce p dimensional data. But a bad choice of the involved constants for computing the random numbers had been made, as exactly 15 hyperplanes are generated for three dimensional data (see Figure 2.17). Obviously this is not a good random generator.

Table 4.1 shows how many planes we have to examine if the distance between two planes is less equal five degrees. If we wanted to revise all planes for a six dimensional dataset to find a structure and we would watch every plane for just one second we would need already 23 days of uninterrupted watching of the screen.

As a consequence we need other methods to pick out the interesting projections which lead to the exploratory projection pursuit.

For a detailled overview about grand tours see Buja, Cook, Asimov & Hurley (1996).

Dimension	No. of planes
3	263
4	51684
5	$\sim$ 9000000
6	$\sim$ 2000000000
7	$\sim$ 200000000000
8	$\sim$ 40000000000000
9	$\sim$ 6000000000000000
10	$\sim$ 800000000000000000
12	$\sim$ 2000000000000000000000
14	$\sim$ 4000000000000000000000000
16	$\sim$ 7000000000000000000000000000
20	$\sim$ 3000000000000000000000000000000000

TABLE 4.1. Number of planes we have to look at if the distance between two planes is less equal five degrees.

4.1.4 Multidimensional Scaling

The aim of multidimensional scaling (MDS) is to find a low dimensional ($d = 1, 2, 3$) space so that the distances d_{rs} between the objects r and s in this space match as close as possible the original dissimilarities δ_{rs} of a higher dimensional configuration space. For an analysis see Cox & Cox (1994).

Several MDS models will be examined: Classical (metric) scaling, least squares scaling, nonmetric scaling.

In metric scaling the dissimilarities δ_{rs} are taken immediately as euclidean distances. An easy algorithm is given by

1. Compute $A = (a_{rs}) = (-0.5\delta_{rs}^2)$
2. Compute $B = (a_{rs} - a_{r.} - a_{.s} + a_{..})$ with $a_{r.}$ the column sums, $a_{.s}$ the row sums and $a_{..}$ the total sum of A
3. Find the eigenvalues λ_i and the eigenvectors v_i of B and normalize so that $v_i^T v_i = \lambda_i$
4. Take the first eigenvectors corresponding to the largest eigenvalues and show them in a plot

The algorithm shows one weakness: we have to choose the dimension of the projection. If the dissimilarities come from euclidean distances then B is a

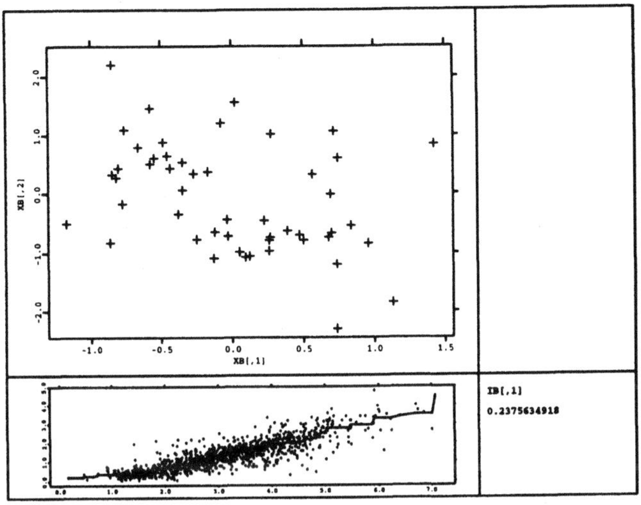

FIGURE 4.2. Nonmetric multidimensional scaling on a subset of the Swiss banknote data (each fourth observation of the data is included).

positive semidefinite matrix. If B is a positive semidefinite matrix it follows that the eigenvalues are positive or zero. Thus a first choice would be to take the number of the nonzero eigenvalues as dimension d .

Since it holds that

$$\frac{1}{2}\sum_{r=1}^{n}\sum_{s=1}^{n} d_{rs}^2 = n\sum_{i=1}^{n-1}\lambda_i$$

we can use the proportion of the variance as a measure, explained by using d dimensions

$$\frac{\sum_{i=1}^{d}\lambda_i}{\sum_{i=1}^{n}\lambda_i}$$

if B is positive semidefinite. If B is not positive semidefinite

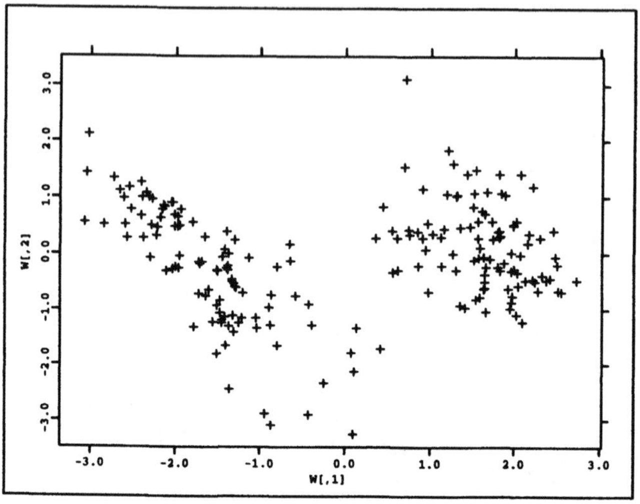

FIGURE 4.3. Metric multidimensional scaling on the Swiss banknote data with euclidean distances. Table 4.2 shows that we recover the dimensionality of the dataset exactly.

$$\frac{\sum_{i=1}^{d} \lambda_i}{\sum_{i=1}^{n} |\lambda_i|}.$$

then the standard methods of PCA can be used to choose the number of dimensions.

In least squares scaling a monotone (parametric) transformation f of the dissimilarities δ_{rs} is added. To find a good low dimensional projection the (stress) functional

$$S_1 = \frac{\sum_{r \neq s} w_{rs}(d_{rs} - f(\delta_{rs}))^2}{\sum_{r \neq s} d_{rs}^2}$$

is minimized (w_{rs} being appropriate weights).

Nonmetric scaling assumes that the level of measurement is at the nominal or, at best, at the ordinal scale. The transformation function f is a monotone

i	λ_i
1	597.061
2	186.188
3	48.439
4	38.737
5	16.957
6	7.067
7	4.743e-013
8	5.247e-014
9	4.105e-014
10	3.710e-014
	...
200	-3.167e-013

TABLE 4.2. We see the eigenvalues computed by MDS. The eigenvalues larger than six are zero, only rounding effects make then different from zero. We discover that the Swiss banknote dataset is six dimensional (as expected).

function such that

$$f(d_{rs}) \leq f(d_{tu}) \text{ if } \delta_{rs} < \delta_{tu}$$

which means that the dissimilarity influences the stress function only indirectly

$$S_2 = \left(\frac{\sum_{r,s} (d_{rs} - f(d_{rs}))^2}{\sum_{r,s} d_{rs}^2} \right)^{1/2}.$$

The stress function S_2 and the minimization of it was proposed in Kruskal (1964a, 1964b). The algorithm is

1. Choose an initial configuration X
2. Normalize the configuration so that mean(X) = 0 and var(X) = 1
3. Compute d_{rs}
4. Fit $f(d_{rs})$, e.g. by monotonic least squares regression
5. Compute a new configuration X by minimizing the stress function
6. Go to 2.

4.2 The Basis of Exploratory Projection Pursuit

Diaconis & Freedman (1984) gave theorems that ***show that under suitable conditions, most projections are approximately gaussian.*** This emphasizes the search for projections where the data are not distributed normally. Thus we put an index value on each projection , which describes the departure from nonnormality. If we try to maximize this index function, we will end up with some most nonnormal projections. This method is called exploratory projection pursuit. The term "projection" implies looking at projected data and the term "pursuit" finding a "good" projection for the purpose of the analysis. In fact this reduces the amount of the projections we have to go through. But we should not see EPP only as an extension of the grand tour. In practice the combination of both showed very fruitful results (Cook, Buja, Cabrera & Hurley 1995).

The idea of exploratory projection pursuit was introduced by Kruskal (1969, 1972). The approach was first successfully implemented for exploratory purposes by Friedman & Tukey (1974). Alternative projection indices have been proposed among others by Jee (1985), Huber (1985), Jones & Sibson (1987), Friedman (1987), Hall (1989a), Cook & Cabrera (1992), Cook, Buja & Cabrera (1993) and Posse (1995a). The idea has been applied to regression analysis (Friedman & Stuetzle 1981b), density estimation (Friedman, Stuetzle & Schroeder 1984), classification (Friedman & Stuetzle 1981a) and discriminant analysis (Polzehl 1995). For projection pursuit regression the approximation of the regression function is characterized in Donoho & Johnstone (1989), convergence rates are obtained in Hall (1989b). Good references about projection pursuit are Jones & Sibson (1987) and Huber (1985).

4.2.1 Projection Pursuit Indices

We want to describe some common projection pursuit indices as they are implemented in `XGobi` for example. The first successful implementation of an index was done by Friedman & Tukey (1974). Their index was based on heuristic arguments. The index

$$I_{FT}(\alpha) = s(\alpha)d(\alpha)$$

is composed of two parts, one s depending only on the covariance structure and one d which captures the "local clusters" of the data. The term s can be avoided if the p dimensional data X were standardized, i.e. $E(X) = 0$ and $Cov(X) = I_p$. In Jones & Sibson (1987) d is expanded as kernel density estimate

$$\hat{f}_Y(y) = \frac{1}{nh} \sum_{j=1}^{n} K\left(\frac{y - y_j}{h}\right)$$

with $Y = \alpha^T X$, $y_i = \alpha^T X_i$ and h a bandwidth. The E could be written as

$$\hat{I}_{FT}(\alpha) = \frac{1}{n^2 h} \sum_{i=1}^{n} \sum_{j=1}^{n} K\left(\frac{y_i - y_j}{h}\right).$$

It turns out that this index is an estimate of

$$I_{FT}(\alpha) = \int_{\mathbb{R}} f_Y^2(y)dy = E_Y(f_Y(y)),$$

which is minimized by a parabolic density if X is standardized. The parabolic density is close to a standard normal density, thus a departure of a parabolic density is also a departure from the standard normal density. Huber (1985) proposed an index based on the negative entropy, which is minimized by the standard normal density:

$$I_E(\alpha) = \int_{\mathbb{R}} f_Y(y) \log(f_Y(y))dy = E_Y(\log(f_Y(y))).$$

Again an estimate is obtained by

$$\hat{I}_E(\alpha) = \frac{1}{n} \sum_{i=1}^{n} \log\left(\frac{1}{nh} \sum_{j=1}^{n} K\left(\frac{y_i - y_j}{h}\right)\right).$$

An extension to multivariate indices can easily be done by extending the kernel K to a multivariate kernel. Common kernels are given in Table 4.3.

The estimates will change slightly to

$$\begin{aligned} \hat{I}_{FT}(\alpha_1, ..., \alpha_d) &= \frac{1}{n^2 h^d} \sum_{i=1}^{n} \sum_{j=1}^{n} K_d\left(\frac{\mathbf{y}_i - \mathbf{y}_j}{h}\right) \\ \hat{I}_E(\alpha_1, ..., \alpha_d) &= \frac{1}{n} \sum_{i=1}^{n} \log\left(\frac{1}{nh^d} \sum_{j=1}^{n} K_d\left(\frac{\mathbf{y}_i - \mathbf{y}_j}{h}\right)\right). \end{aligned}$$

Here $\mathbf{y_i}$ denotes $(y_{i1}, ..., y_{id}) = (\alpha_1^T X_i, ..., \alpha_d^T X_i)$ the multivariate projection. We assume that the projection vectors $\alpha_1, ..., \alpha_d$ are orthonormal.

$K_d(x_1, ..., x_d)$	=	$1/c_d$	$I(r<1)$	Uniform
$K_d(x_1, ..., x_d)$	=	$2/c_d(1-r)$	$I(r<1)$	Triangle
$K_d(x_1, ..., x_d)$	=	$1.5/c_d(1-r^2)$	$I(r<1)$	Epanechnikov
$K_d(x_1, ..., x_d)$	=	$15/(8c_d)(1-r^2)^2$	$I(r<1)$	Quartic
$K_d(x_1, ..., x_d)$	=	$30/(13c_d)(1-r^2)^3$	$I(r<1)$	Triweight
$K_d(x_1, ..., x_d)$	=	$\pi/(2c_d)\cos(\pi r/2)$	$I(r<1)$	Cosine
$K_d(x_1, ..., x_d)$	=	$(2\pi)^{-d/2}\exp(-r^2/2)$		Gaussian

TABLE 4.3. Common kernels with $r = \sqrt{\sum_{i=1}^d x_i^2}$ and c_d the volume of the d dimensional unit sphere.

Of course the negative entropy is not the only functional which minimizes the standard normal density. Jee (1985) proposed to use the Fisher information as index

$$I_{FI} = \int_{\mathbb{R}} \frac{(f'_Y(y))^2}{f_Y(y)} dy.$$

But this estimate does not involve only the evaluation of the density, but that of the derivative as well. This makes the index complicated to use.

Cook & Cabrera (1992) build up two indices based on two estimations of $d(X)$. Both indices show for the RANDU data a deep minimum for the projection in Figure 2.17.

Posse (1995a) uses a tesselation of the $\mathbb{R}^2$ plane in 48 bins (see Figure 4.24). The index is defined by

$$I_P(\alpha, \beta) = \frac{4}{\pi}\int_0^{\pi/4}\sum_{i=1}^{4} 8\frac{\left(\int_{B_k} f_Y(y,\eta) - \phi(y)dy\right)^2}{\int_{B_k}\phi(y)dy}$$

with $f_Y(y,\eta)$ the empirical density after a rotation of an angle η and estimated by

$$\hat{I}_P(\alpha, \beta) = \frac{1}{ln^2}\sum_{j=0}^{l-1}\sum_{k=1}^{48}\frac{\left(\sum_{i=1}^{n} I(Y_i(\eta_j) \in B_k) - c_k n\right)^3}{c_k}$$

with $\eta_j = \pi j/(4l)$ and $Y_i(\eta_j)$ being the rotated X_i in the projection plane. Huber claims that the χ^2 distances behave mostly like the human eye as to pattern recognition (personal communication with Posse 1990).

Yenjukov (1989) builds up an index from the weighted angles. He constructs an angular index from

$$I_{YA}(\alpha, \beta) = \int_{-\pi}^{\pi} (f_\eta(\eta) - 1/(2\pi))^2 d\eta$$

with η the angle. Since the harmonic moments of the uniform distribution are zero we get an estimate by

$$\hat{I}_{YA}(\alpha, \beta) = \sum_{j=1}^{m} \hat{a}_j^2 + \hat{b}_j^2$$

with $\hat{a}_j = 1/n \sum_{k=1}^{n} \sin(i\eta_i)$ and with $\hat{b}_j = 1/n \sum_{k=1}^{n} \cos(i\eta_i)$.

The drawback is that this index does not take the radial information in account. Thus he defined the coefficients

$$\begin{aligned} a_j &= E(h(r)\sin(i\eta)) \\ b_j &= E(h(r)\cos(i\eta)) \end{aligned}$$

with $h(r)$ a transformation of the radial information, e.g. $h(r) = r^q$, $h(r) = r^q/\sqrt{(}Dr^q)$ or $h(r) = r^q/(c + dr^q)$.

Jones & Sibson (1987) proposed an index based on moments, which approximates the entropy index. Under certain conditions the entropy index can be written as

$$\begin{aligned} \int_{I\!R} f_Y(y)\log(f_Y(y))dy &\approx 0.5 \int_{I\!R} \phi(y)\epsilon^2(y)dy \\ &\approx \frac{4\kappa_3^2 + \kappa_4^2}{48} \end{aligned} \tag{4.2}$$

with $\kappa_3 = \mu_3$ and $\kappa_4 = \mu_4 - 3$ the cumulants and μ_3 and μ_4 the (mixed) central moments. The first moments are known, since the data are standardized ($\mu_1 = 0$, $\mu_2 = I_2$). It is easy to obtain a bivariate index:

$$\hat{I}_M(\alpha_1, \alpha_2) = \frac{4(\kappa_{30}^2 + 3\kappa_{21}^2 + 3\kappa_{12} + \kappa_{03}^2) + \kappa_{40}^2 + 4\kappa_{31}^2 + 6\kappa_{22}^2 + 4\kappa_{13}^2 + \kappa_{04}^2}{48}$$

with $\kappa_{rs} = \mu_{rs}$ for $r+s=3$ and $\kappa_{40} = \mu_{40} - 3$, $\kappa_{04} = \mu_{04} - 3$, $\kappa_{13} = \mu_{13}$, $\kappa_{31} = \mu_{31}$ and $\kappa_{22} = \mu_{22} - 1$. The index measures mainly departure from skewness and kurtosis. The advantage in the calculation is that the index does not depend upon the number of observations.

All of the next three indices are based on orthonormal function expansion, and it is assumed that the data are standardized. The first index was developed by Friedman (1987). The idea of Friedman was to capture clusters in the data. He transformed the projected data by the standard normal cumulative distribution function

$$z = 2\Phi(y) - 1, \tag{4.3}$$

such that Z is uniformly distributed if Y is standard normally distributed. The index is defined by

$$\begin{aligned} I_L(\alpha) &= \int_{-1}^{1} (f_Z(z) - 0.5)^2 dz \\ &= \int_{-1}^{1} f_Z^2(z) dz - 0.25. \end{aligned}$$

A multivariate formula can be obtained by transforming each variable Y_j to Z_j as in 4.3. Thus an estimate for the Legendre index is

$$I_L(\alpha_1, ..., \alpha_d) = \frac{1}{n^2} \sum_{i_1=0}^{i_1+...+i_d \leq J} \cdots \sum_{i_d=0} \left(\sum_{j=1}^{n} P_{i_1}(z_{j1}) ... P_{i_d}(z_{jd}) \right)^2 .$$

Following Cook et al. (1993) this index can be rewritten as

$$I_L(\alpha_1, ..., \alpha_d) = \int_{\mathbb{R}^d} (f_Y(\mathbf{y}) - \phi(\mathbf{y}))^2 \frac{1}{2^d \phi(\mathbf{y})} d\mathbf{y},$$

which shows clearly that the differences in the tails of the distribution are weighted much stronger than the differences in the center distribution. In practice this index is attracted by skewed distributions. Concerned about theoretical properties of the Legendre index Hall (1989a) developed a new index, the Hermite index, based on Hermite polynomials

$$I_H(\alpha_1, ..., \alpha_d) = \int_{\mathbb{R}^d} (f_Y(\mathbf{y}) - \phi(\mathbf{y}))^2 d\mathbf{y}.$$

We get an estimate by

$$\hat{a}_{k_1 \dots k_d} = \frac{1}{n} \sum_{i=1}^{n} H_{k_1}(y_{i1}) \dots H_{k_d}(y_{id}),$$

and we can estimate the integral by

$$\hat{I}_L = (\hat{a}_{0\dots 0} - 1)^2 + \sum_{i_1=0}^{1 \le i_1 + \dots + i_d \le J} \dots \sum_{i_d=0} \hat{a}^2_{i_1 \dots i_d}.$$

Cook et al. (1993) developed a new index, the Natural-Hermite index, to come back to Friedman's original idea of upweighting the differences in the center of the distribution. They defined their index via

$$I_{NH}(\alpha_1, \dots, \alpha_d) = \int_{\mathbb{R}^d} (f_Y(\mathbf{y}) - \phi(\mathbf{y}))^2 \phi(\mathbf{y}) d\mathbf{y}.$$

The integral simplifies with the use of an orthonormal function system to

$$\begin{aligned}
I_{NH}(\alpha_1, \dots, \alpha_d) &= \int_{\mathbb{R}^d} (f_Y(\mathbf{y}) - \phi(\mathbf{y}))^2 \phi(\mathbf{y}) d\mathbf{y} \\
&= \int_{\mathbb{R}^d} \Bigg(\sum_{i_1=0}^{\infty} \dots \sum_{i_d=0}^{\infty} (a_{i_1 \dots i_d} - b_{i_1 \dots i_d}) \\
&\qquad\qquad H_{e_{i_1}}(y_1) \dots H_{e_{i_d}}(y_d) \Bigg)^2 d\mathbf{y} \\
&= \sum_{i_1=0}^{\infty} \dots \sum_{i_d=0}^{\infty} \sum_{j_1=0}^{\infty} \dots \sum_{j_d=0}^{\infty} (a_{i_1 \dots i_d} - b_{i_1 \dots i_d})(a_{j_1 \dots j_d} - b_{j_1 \dots j_d}) \\
&\qquad \int_{\mathbb{R}^d} H_{e_{i_1}}(y_1) \dots H_{e_{i_d}}(y_d) H_{e_{j_1}}(y_1) \dots H_{e_{j_d}}(y_d) \phi(\mathbf{y}) d\mathbf{y} \\
&= \sum_{i_1=0}^{\infty} \dots \sum_{i_d=0}^{\infty} (a_{i_1 \dots i_d} - b_{i_1 \dots i_d})^2
\end{aligned}$$

The coefficient $a_{k_1 \dots k_d}$ can be estimated by

$$\hat{a}_{k_1 \dots k_d} = \frac{1}{n} \sum_{i=1}^{n} H_{e_{k_1}}(y_{i1}) \dots H_{e_{k_d}}(y_{id})$$

whereas the coefficients $b_{k_1 \dots k_d}$ can be calculated analytically from Abramowitz & Stegun (1972)

$$\begin{aligned} b_{i_1 \ldots i_d} &= b_{i_1} \ldots b_{i_d} \\ b_k &= \begin{cases} 0 & \text{if } k \text{ is odd} \\ \frac{(-1)^{k/2}\sqrt{k!}}{\sqrt{\pi}(k/2)!2^{k+1}} & \text{if } k \text{ is even} \end{cases} . \end{aligned}$$

Thus an estimate is given by

$$\hat{I}_{NH}(\alpha_1, ..., \alpha_d) = \sum_{i_1=0}^{i_1+...+i_d \leq J} \cdots \sum_{i_d=0} (\hat{a}_{i_1 \ldots i_d} - b_{i_1 \ldots i_d})^2$$

4.2.2 *Relation of Other Techniques to EPP*

After the introduction of several index functions we can easily establish a relation to the previously mentioned techniques:

- Principal component analysis
  If we introduce an index function of the form

  $$I(\alpha_1, ..., \alpha_d) = \sum_{i=1}^{d} \text{Var}(\alpha_i^T X)$$

  and maximize it, we find the first d principal components.

- Andrew's curves
  The main relationship is that we can see one special univariate projection of a multivariate dataset for each t.

- Grand tour
  The grand tour can be regarded as an unguided EPP method, that means

  $$I_{GT}(\alpha_1, ..., \alpha_d) = 0.$$

  No (multivariate) projection is preferred.

- Multidimensional scaling
  In least squares scaling and nonmetric scaling the stress function S_1 and S_2 will be minimized. If we define the stress index

  $$I_1(\alpha_1, ..., \alpha_d) = -S_1 \text{ and } I_2(\alpha_1, ..., \alpha_d) = -S_2$$

  we get a projection pursuit index which has to be maximized.

4.2.3 *The Index Functions in Practice*

For our explanations we will use two three dimensional artificial datasets. The first dataset (CYL-data) contains 100 random points located on the surface of a cylinder with both length and radius equal 1.

The second dataset (TET-data) consists of 250 points located on the supporting hyperplanes of a tetrahedron given by its facets. Three hyperplanes contain 50 normally distributed random points each while the fourth hyperplane accommodates 100 normally distributed random points.

In case of the CYL-data we find a situation with one striking two dimensional projection while in case of the TET-data there are six interesting views determined by the pairs of the supporting hyperplanes. The latter dataset will be used to demonstrate interesting effects concerning the choice of bandwidths and polynomial orders. The distinguished projections of the datasets are displayed in Figure (4.4) and Figure (4.5). The lower row of Figure (4.5) contains the projection determined by the intersection of the basis plane with one of the others. The upper row contains projections determined by pairs of the first three planes, i.e. the structure displayed contains 100 points in case of the upper row and 150 in case of the lower.

For every two dimensional projection of a three dimensional data set, the projection plane can be identified uniquely by the normal vector. The normal vector can be expressed in terms of two angles φ and ϑ. The interesting projections correspond to the following values of φ and ϑ:

$$\begin{array}{lll} \varphi=-1.27, \vartheta=0.55 & \varphi=-1.2, \vartheta=-0.13 & \varphi=-0.56, \vartheta=0.22 \\ \varphi=-0.79, \vartheta=-1.35 & \varphi=0.53, \vartheta=-0.49 & \varphi=0.77, \vartheta=0.53 \end{array}$$

The uniform kernel leads to noncontinuous piecewise constant indices while triangle and epanechnikov kernels provide continuity but no differentiability. Because of the maximization of the index function involved in the projection pursuit approach the uniform kernel is not helpful although being easy to compute, while triangle and epanechnikov kernels do not provide the wanted qualitative properties of the indices.

We restricted ourselves to the triweight kernel which secures the existence of second derivatives of the indices and which is much easier to compute than the alternative gaussian kernel. The kernel has limited support, a circle with a radius equal to the selected bandwidth.

The essential question in exploratory projection pursuit seems to select an appropriate bandwidth. For a small bandwidth we would expect the index function to have numerous local maxima corresponding to small clusters of observations in the projection plane.

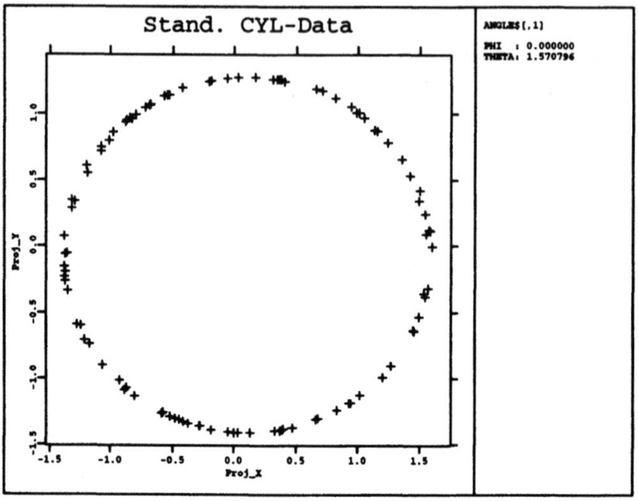

FIGURE 4.4. Optimal projection of CYL-data

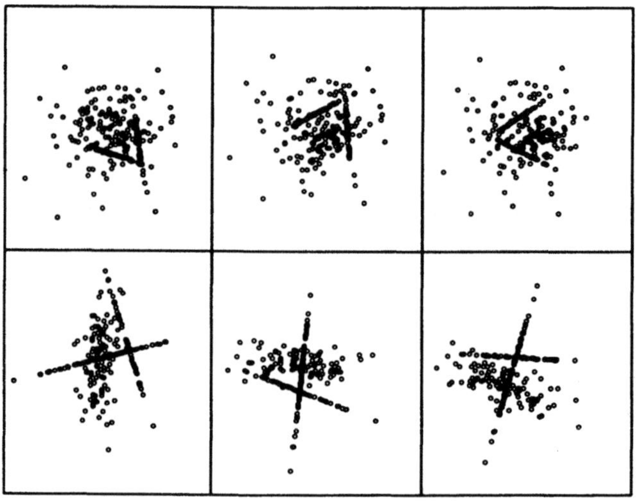

FIGURE 4.5. Optimal projection of TET-data

If the bandwidth increases the number of maxima will decrease as the local maxima will melt together. Further increase of the bandwidth will lead the index function to show just a few, maybe only one local maximum.

Finding one global maximum is a complicated task in case of many local maxima. In exploratory projection pursuit the aim usually is not to pick up the global maximum but to identify a set of distinct local maxima. Projections corresponding to a local maximum provide an "interesting" view to the data, too.

Silverman (1986) gives the following formulas for an optimal bandwidth minimizing the asymptotic mean integrated squared error (AMISE). If the kernel is a radially symmetric probability density function and the unknown density is bounded and has continuous second derivatives, the AMISE for a given h results in

$$AMISE = \frac{\int_{\mathbb{R}^d} K^2(t)dt}{nh^d} + \frac{1}{4}h^4 \left(\int_{\mathbb{R}^d} t_1^2 K(t)dt\right)^2 \left(\int_{\mathbb{R}^d} tr(\nabla^2 f)^2(t)dt\right)$$

which leads to

$$h_{opt}^{d+4} = dn^{-1} \left(\int_{\mathbb{R}^d} K^2(t)dt\right) \left(\int_{\mathbb{R}^d} t_1^2 K(t)dt\right)^{-2} \left(\int_{\mathbb{R}^d} tr(\nabla^2 f)^2(t)dt\right)^{-1},$$

(see Scott (1992), 6.48) where d stands for the dimensionality of the projected data. Analog to the "rule-of-thumb" we can plug in the multivariate normal density instead of f, which simplifies the last term to

$$\int_{\mathbb{R}^d} tr(\nabla^2 \phi)^2(t)dt = (4\pi)^{d/2}(d/2 + d^2/4).$$

Plugging-in the data of our example, the "rule-of-thumb" reference bandwidth becomes:

$$h_{rot}^{d+4} = \frac{d}{n} R(K) \left(\int_{\mathbb{R}^d} t_1^2 K(t)dt\right)^{-2} (4\pi)^{-d/2} \frac{4}{d(d+2)}. \tag{4.4}$$

and plugging in the triweight kernel

$$h_{rot} = 3.12n^{-1/6}$$

which leads to a "rule-of-thumb" bandwidth $h_{rot} \approx 1.45$ in case of the CYL-data and to $h_{rot} \approx 1.24$ in case of the tetrahedron data.

The polynomial based indices under consideration measure differences between the underlying density of the projected data and a standard normal density. In order to detect differences which are not caused by location and covariance structure the data should be standardized in a first step.

Even small deviations from a standardized situation lead to strong changes of the index functions. The effect can be regarded as a mapping of information by location and covariance effects.

The problem of bandwidth selection for kernel indices relates with the problem of specification of an appropriate order J of the orthogonal series approximations in case of the polynomial indices. Selection of a small order usually corresponds to a consideration of global effects, while a high order will allow a more local modeling of the structure contained in the data. The situation is similar to case of kernel indices in that sense that a high order J will cause numerous local maxima of the index function while a small J will correspond to only a few but dim maxima.

A very unpleasant effect occurs when using the Legendre index which turns out to be not invariant with respect to rotation inside the projection plane. This effect is avoided by use of rotation symmetric kernels in case of kernel indices.

Figures 4.6 - 4.10 illustrate the behaviour of the index functions for the CYL-Data. The figures show contour plots of the index functions computed on a net of 53×53 points for (φ, ϑ). In case of the polynomial indices the maximum was used with respect to rotation inside the projection plane. The levels used correspond to the $0.5, 0.67, 0.8, 0.9, 0.95, 0.975, 0.99$ and 1-quantiles of index values on the net. Figures 4.11 - 4.15 show analogous contour plots for the tetrahedron data.

Figures 4.6 and 4.7 illustrate the expected behaviour of the kernel based indices. Small bandwidths h lead to a huge number of local maxima corresponding to occasional clusters of observations in meaningless projections. Because of the outstanding structure most of the local maxima are small compared with the global maximum but nevertheless they cause tremendous problems in numerical maximization of the index functions.

Bandwidths in the magnitude of the "rule-of-thumb" bandwidth behave well in this example while large bandwidths lead to oversmoothing effects blurring the structure for $h = 1.6$ and hiding the structure completely for $h = 3.2$.

Figures 4.8, 4.9 and 4.10 show the corresponding plots for the polynomial based indices. The Legendre index and, to some extent, the Hermite index show evident problems when handling even this clear structure in case of a small order J.

The Natural-Hermite index behaves much better in this situation as expected in Cook et al. (1993). Even for larger values of J the estimated criteria are sufficiently smooth providing only a small number of local minima.

The situation is much more complicated for the second dataset. Figures 4.11 and 4.12 show contourplots of the Friedman-Tukey and the entropy index. The projections shown in Figure 4.5 are marked by a star. In case of the smallest bandwidth $h = 0.1$ we get local maxima corresponding to all interesting projections, although for this bandwidth there are 141 local maxima on the net in case of the Friedman-Tukey index and 93 local maxima in case of the Entropy index.

Increasing the bandwidth leads to smearing effects still presenting interesting projections of the basis plane though melting together the maxima corresponding to the wanted projections. The "rule-of-thumb" bandwidth is clearly too large to show the underlying structure of a tetrahedron. The maxima corresponding to the projections in the upper row of Figure 4.5 are smoothed away resulting in a local maximum without much information while the basis plane still can be identified. The larger bandwidths $h = 1.6$ and $h = 3.2$ lead to strong oversmoothing effects hiding the structure completely.

The corresponding results for the polynomial indices are displayed in Figure 4.13, 4.14 and 4.15. In case of small order $J \leq 3$ all indices fail completely to find the structure. For $J > 3$ the basis plane is identified. In case of the Legendre index $J = 10$ gives sufficient information about the structure while the Hermite and Natural-Hermite index require a further increase of J. It can be observed that for a medium J a sufficient number of local maxima exists but global information considered in the polynomial approximation leads to a smoothing sufficient enough to hide the structure. A substantial increase of the order J of the approximation leads to longer computing times than needed as in case of the kernel based indices.

The optimal bandwidth h or order of approximation J seems to depend strongly on the underlying structure of the data. We would suggest to use smaller bandwidths and larger values of J rather than "optimal" bandwidths or orders because of hiding effects in case of complicated structures. Using a small h or large J will lead to a considerable amount of local maxima especially in case of the kernel based indices. This will cause a lot of numerical complications, requiring stochastic or deterministic search algorithms combined with a numerical maximization in order to improve accuracy. For simple structures it makes sense to follow an idea of Hall (1989a) to start with larger bandwidths or lower order of approximation in a initial step to identify the maxima and then to decrease the bandwidth or to increase the order for a numerical maximization step, although some interesting projections might be lost in case of complicated structures. Usually it is interesting to determine all local maxima exceeding a certain level of the criteria. The projections corresponding to these maxima have to be inspected visually to evaluate the information contained. This indicates that a stochastic search algorithm concentrating on areas of this high index values could be used.

4.2.4 The "MSE-FT"-optimal Bandwidth

As another choice for a reference bandwidth we can use a bandwidth which is computed from the minimization of the approximated mean integrated

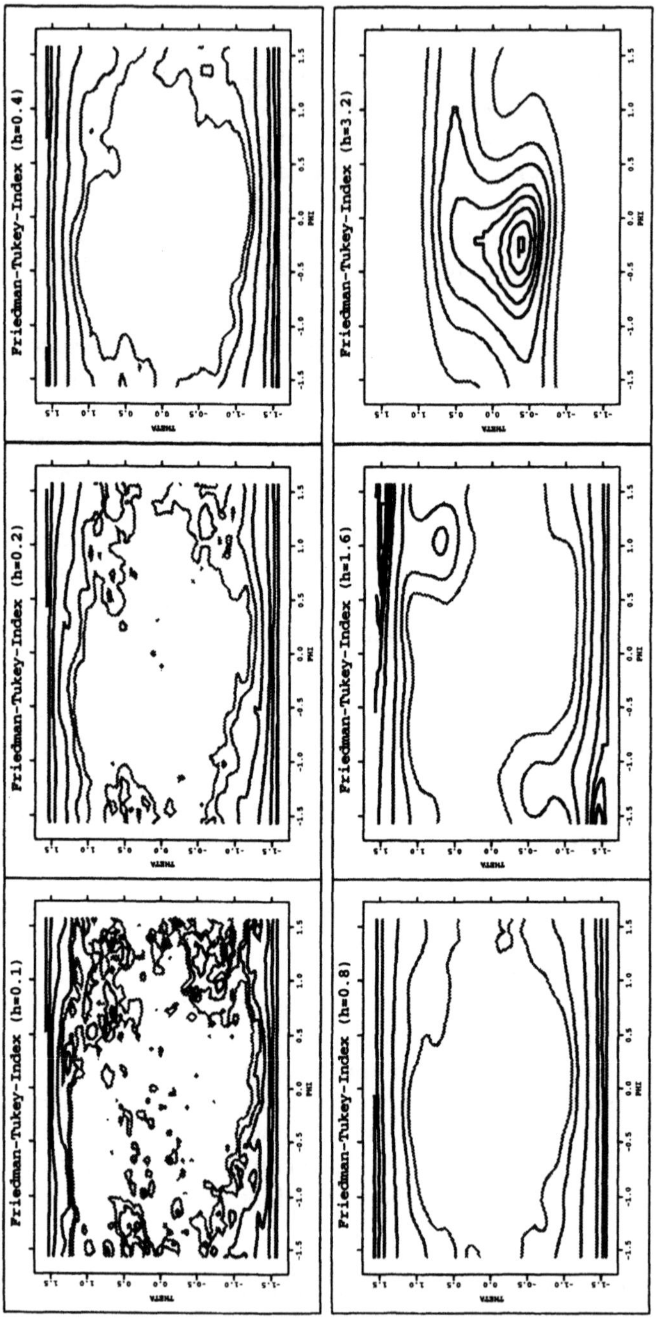

FIGURE 4.6. CYL-Data Friedman-Tukey index for different bandwidths

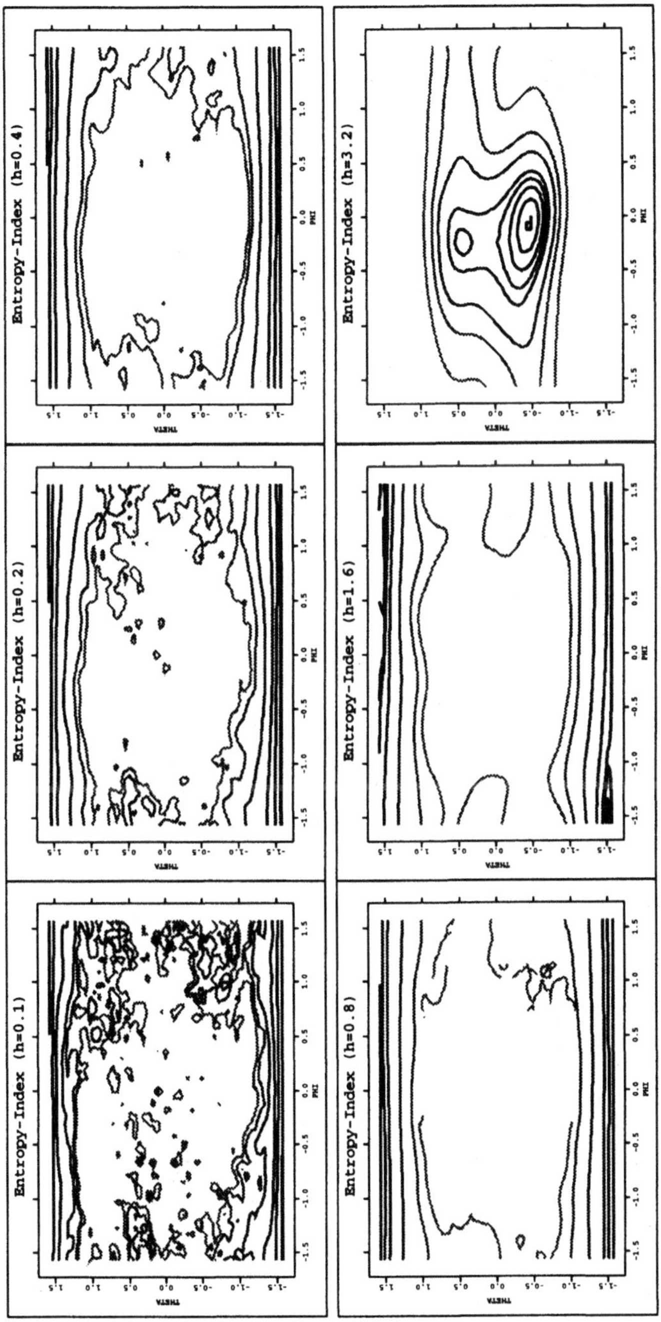

FIGURE 4.7. CYL-Data entropy index for different bandwidths

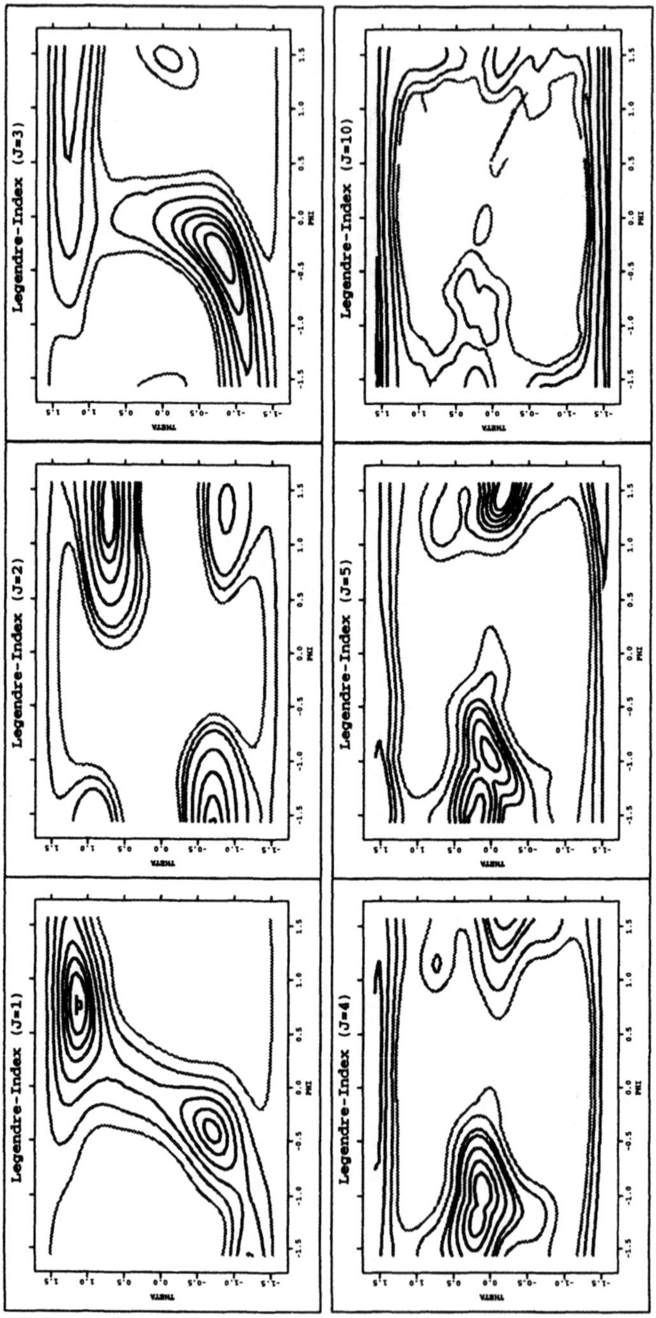

FIGURE 4.8. CYL-Data Legendre index for different order J

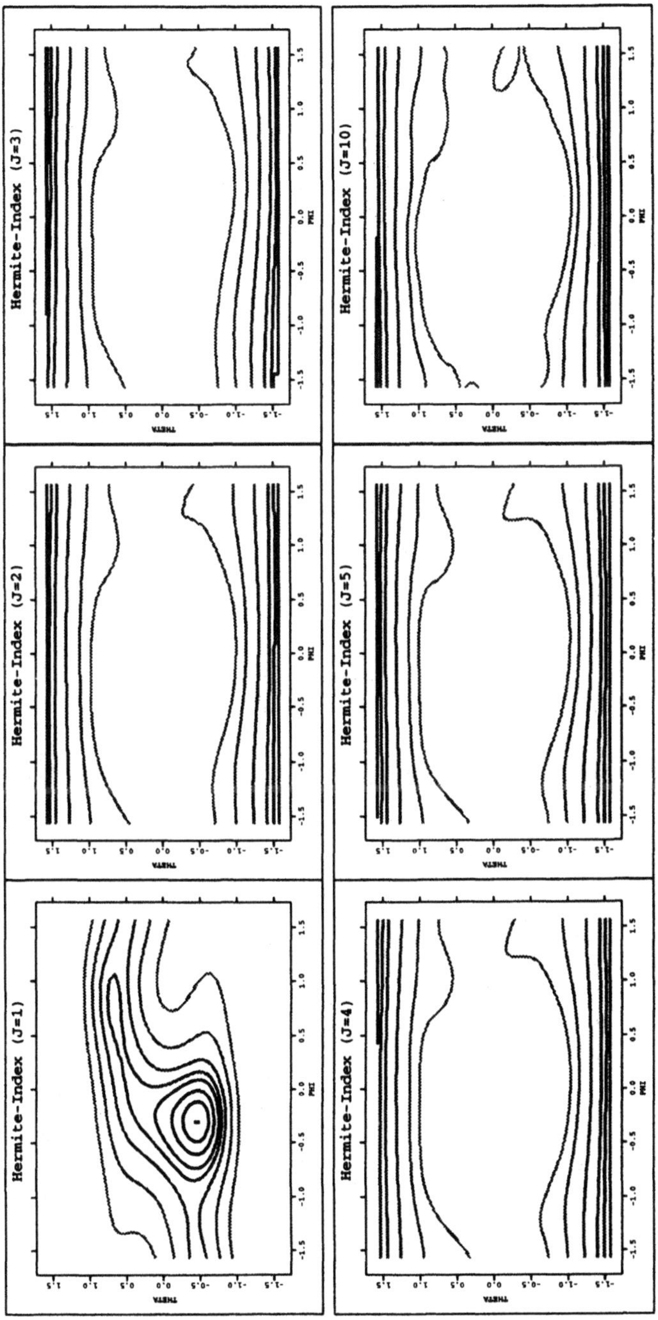

FIGURE 4.9. CYL-Data Hermite index for different order J

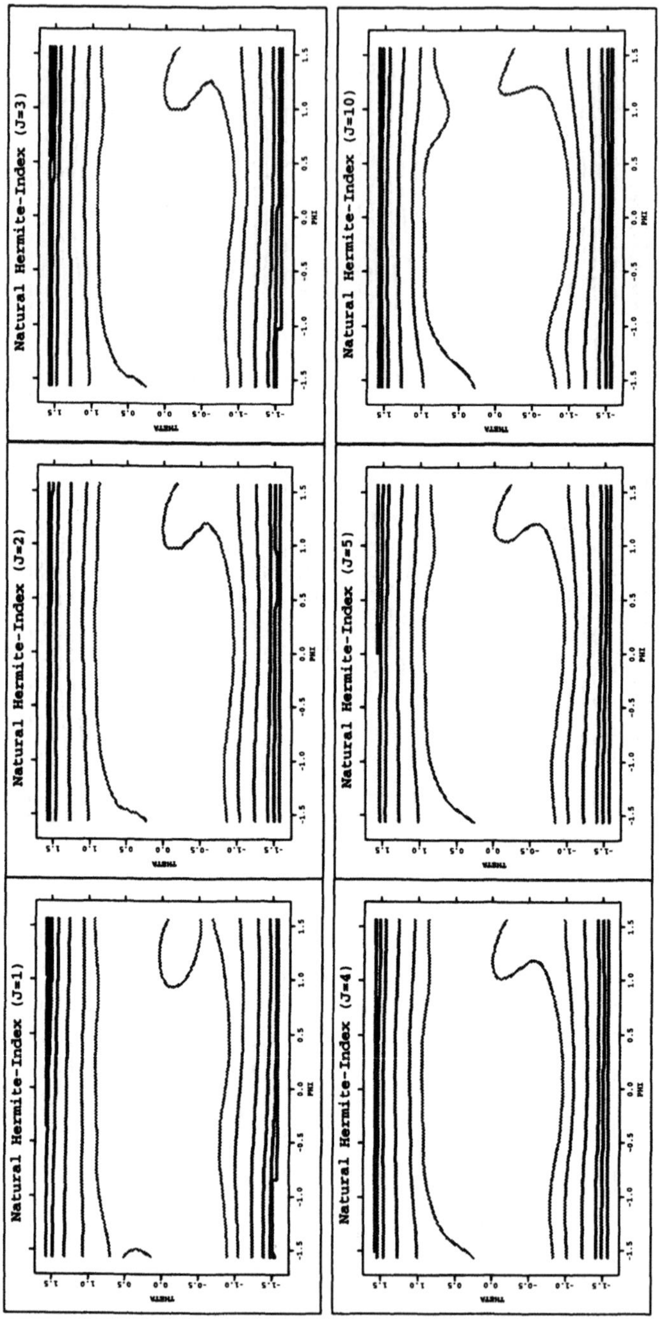

FIGURE 4.10. CYL-Data Natural-Hermite index for different order J

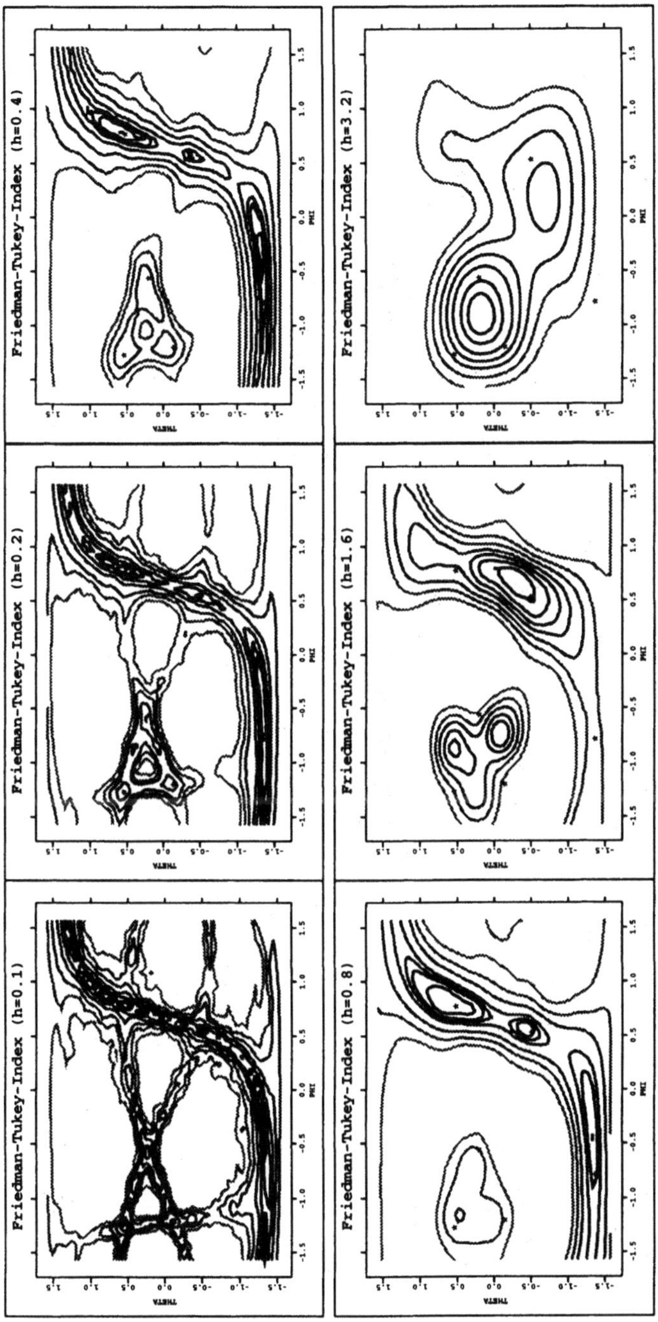

FIGURE 4.11. TET-Data Friedman-Tukey index for different bandwidths

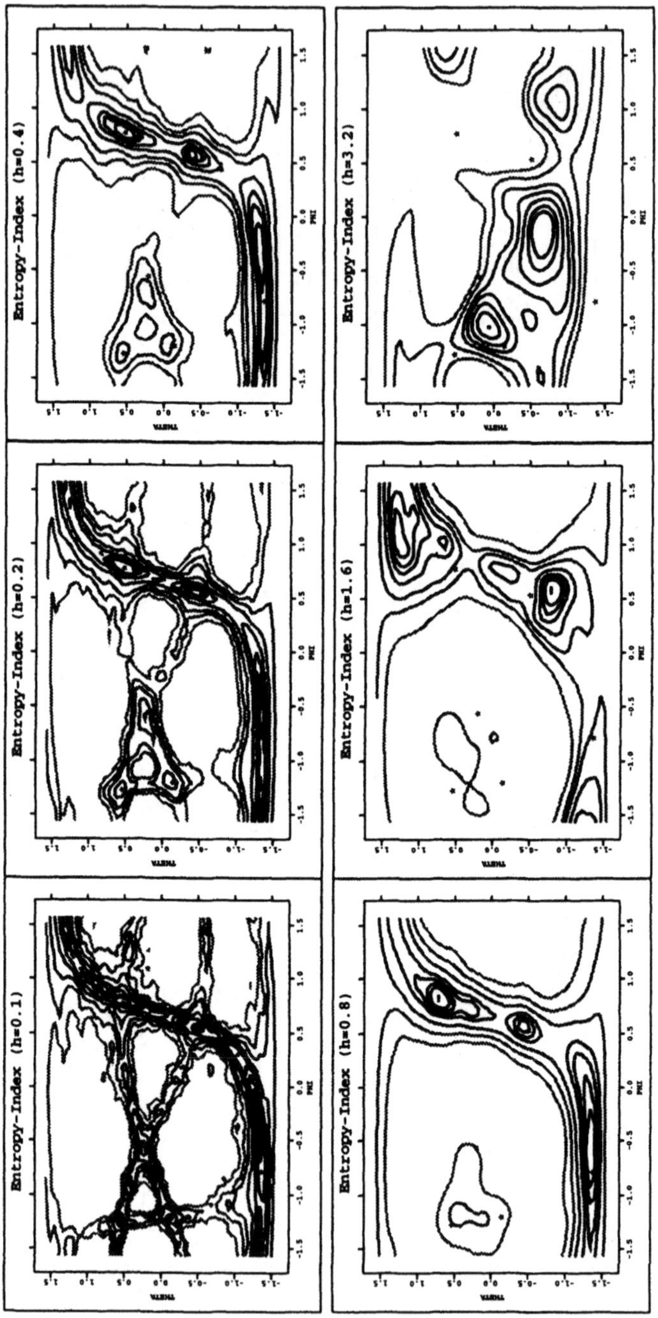

FIGURE 4.12. TET-Data entropy index for different bandwidths

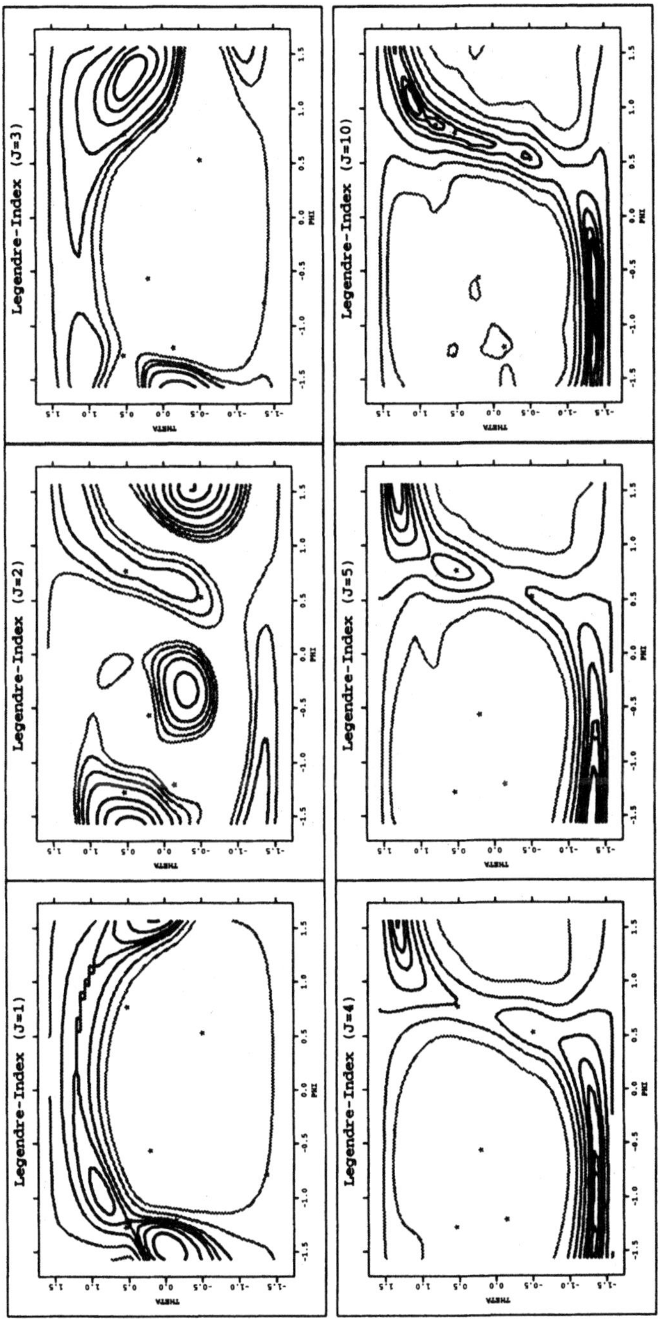

FIGURE 4.13. TET-Data Legendre index for different order J

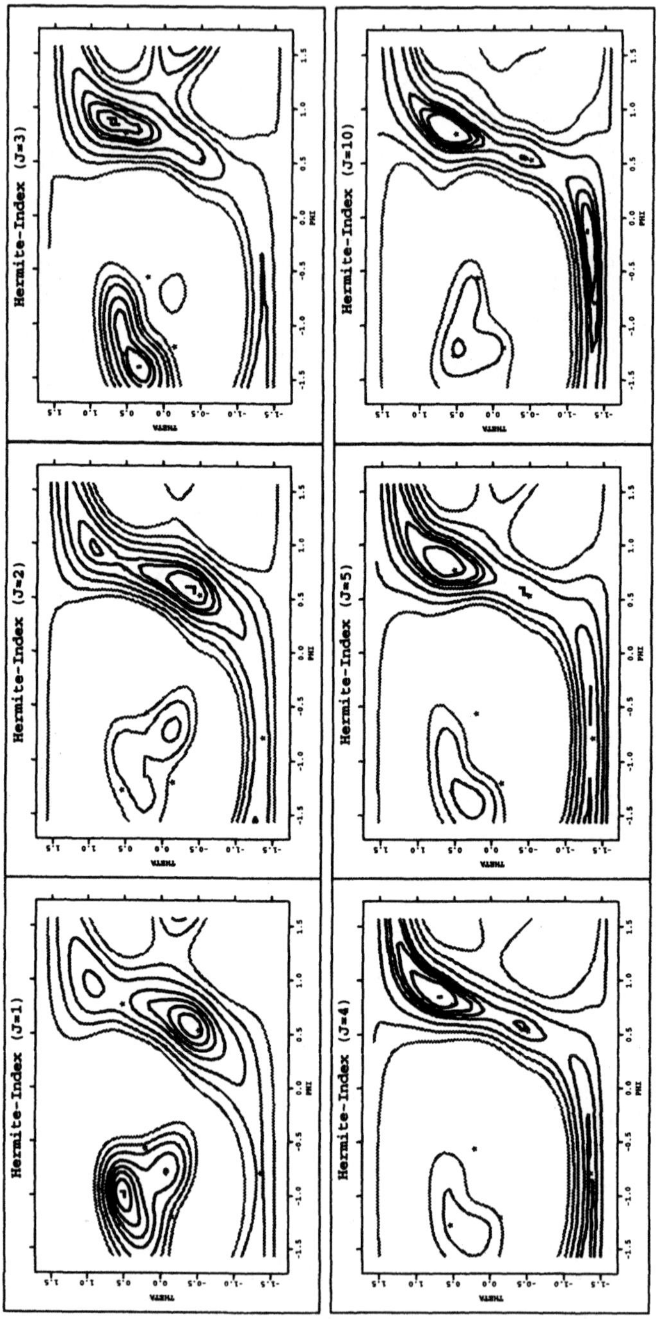

FIGURE 4.14. TET-Data Hermite index for different order J

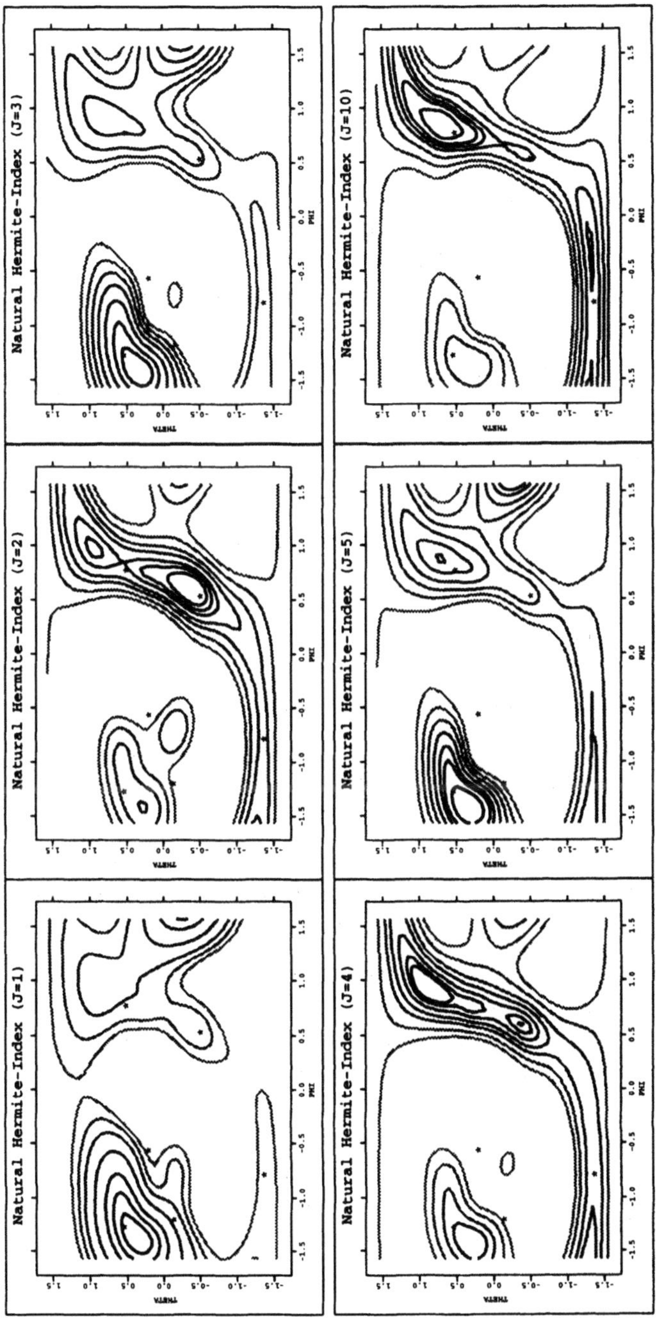

FIGURE 4.15. TET-Data Natural-Hermite index for different order J

squared error. We will see that we can derive a smaller reference bandwidth. It holds for the Friedman-Tukey index

$$MSE(\hat{I}_{FT}) = \frac{C_0}{h^4} + \frac{C_1}{h^2} + C_2 + C_3 h^2 + O(h^4, h^4 n^{-1}, h^4 n^{-2}, ...)$$

with constants $C_0 = O(n^{-2}), C_1 = O(n^{-2}), C_2 = O(1)$ and $C_3 = O(1)$ which can be found in Klinke & Cook (1995). To get an estimate for the "MSE-FT"-optimal bandwidth we drop the $O(h^4)$-term, and to find the minimum for the approximated MSE ($AMSE$) we compute the derivative

$$\frac{d\,AMSE(\hat{I}_{FT})}{d\,h} = \frac{-4C_0}{h^5} + \frac{-2C_1}{h^3} + 2C_3 h = 0$$

and multiply with h^5

$$-4C_0 - 2C_1 h^2 + 2C_3 h^6 = 0.$$

By replacing h^2 by g we get a reduced polynomial of degree 3, which can be solved by the formula of Cardano (Gellert, Küstner, Hellwich & Kästner 1977). The discriminant is always positive and converges to zero ($n \to \infty$). The positive discriminant allows only one unique solution of the equation.

We can compute a minimum by plugging in the triweight kernel and the standard normal gaussian density for the unknown density. As result we get Table 4.4.

The reference bandwidth for the CYL-data is $h_{mse} \approx 0.83$ and for the TET-data $h_{mse} \approx 0.61$. As we see in the Figures 4.6, 4.7, 4.11 and 4.12 the "MSE-FT"-optimal bandwidth is still to large.

In principle the same could be done with the entropy index. But the comparable calculations would be much more complicated. We would have to replace the log-function by its Taylor-expansion. Obviously a linear approximation is not sufficient if n (or h) varies. This can be seen by the following inequality

$$\log\left(\frac{4}{\pi n^2 h^2}\right) \leq \hat{I}_E < \log\left(\frac{4}{\pi n h^2}\right).$$

n	Bandwidth	
	rule-of-thumb	MSE-FT
2	2.77960	2.716855
5	2.38594	2.325666
10	2.12563	1.877361
20	1.89372	1.464121
50	1.62552	1.059059
100	1.44818	0.834545
200	1.29018	0.659888
500	1.10746	0.485088
1000	0.98663	0.384715
2000	0.87899	0.305229
5000	0.75450	0.224842

TABLE 4.4. Bandwidth computed by using the rule-of-thumb and the use of the MSE

4.2.5 Computational Aspects

Kernel based indices

The projection pursuit indices are involving, as a main task, the estimation of a density. With the kernel based indices this will be done with kernel density estimates.

One technique to improve the computational speed of kernel density estimates is binning. If the dimension of the data grows, however, this technique looses more and more of its advantage.

Another idea is to replace float-operations in the density estimation by integer-operations, which are faster. This can be seen in the comparison of Table 4.5 and Table 4.6.

Operation	Cycles
16 bit integer - addition	7
16 bit integer - subtraction	7
16 bit integer - multiplication	12 - 25
16 bit integer - division	22

TABLE 4.5. Execution time in cycles for different mathematical operations in the 80386

The computational speed of the kernels in Table 4.3 differ a lot. For all the

Operation	Cycles
64 bit float - addition	29 - 37
64 bit float - subtraction	28 - 36
64 bit float - multiplication	32 - 57
64 bit float - division	94

TABLE 4.6. Execution time in cycles for different mathematical operations in the 80387

following computations, the Zortech C++ 3.0 - compiler of Symantec Inc. on a PC 486 with 50 MHz was used. Table 4.7 shows the relative computational time for the evaluation of the bivariate kernels in relation to the uniform kernel. Just like many other compilers the Zortech compiler uses an integrated optimizer. In the right column we see the relative computing time when using the optimizer. So we see that, for example, the epanechnikov kernel needs 16% more time to evaluate the kernel values from the same data as the uniform kernel (the data were uniformly distributed in the right upper quarter of the unit circle). If we do not use the optimizer the uniform kernel takes more than 3 times longer to calculate the kernel values.

Kernel	Unopt.	Opt.
Uniform	3.36	1.00
Epanechnikov	4.53	1.16
Quartic	5.19	1.35
Triweight	5.88	1.41
Triangle	7.19	1.77
Cosine	11.81	5.78

TABLE 4.7. Relative computational time of bivariate kernels

We can distinguish two classes of kernels independent from using unoptimized code (286-code, large memory model, no optimization) or optimized code (386-code, extender, fully time-optimized, using the coprocessor). On the one side we have the polynomial kernels (uniform, quartic, epanechnikov, triangle and triweight), on the other side the transcendental kernels (cosine). On the average the polynomial kernels are 5-7 times faster to calculate than the transcendental kernels.

This also indicates that we should use the optimizer to speed up the programs, which was done for all programs. An optimizer program is able to do all simple optimizations, so that further speed improvements can only come from improvements of the technique being used. Such improvements will be presented in the following.

The first kind of optimization we can do for the calculation of the density at all points X_i is to use the symmetry of the kernel :

$$K_2\left(\frac{X_i - X_j}{h}\right) = K_2\left(\frac{X_j - X_i}{h}\right).$$

The second optimization is, since we know we have to calculate $K_2(0)$ for every datapoint, to calculate this kernel value once.

We now take advantage of the fact that the support of all kernels mentioned in Table 4.3 is the unit circle. We can now continue as described in Silverman (1986), but the closed interval $[1, -1]$ has to be replaced by the compact unit circle. One reason to expel the gaussian kernel was the infinite support ($\mathbb{R}^2$). If we sort the data by the first component, we only have to run from a datapoint with index idx_{low} to an index idx_{high}. The pictures in Figure 4.16 show the relative computational time for using unsorted and sorted data for different kernels (uniform, quartic) and data sets (UNIFORM, LINE, CIRCLE). The data all lie in $[0, 1]^2$, for a more detailed description see section A.4. In Figure 4.16, as in all later figures of this type, the x-axis shows the common logarithm of the sample size. On the y-axis we see the common logarithm of the bandwidth. The graphic shows the ratio of the time for calculating the density with sorted data and unsorted data depending on the bandwidth and the sample size for different kernels and data sets. The thick line always indicates a ratio of 1, which means that both programs need the same time to calculate the density. The lines under the thick line are indicating ratios of 0.8, 0.6, ..., which means that the program with sorted data needs only 80%, 60%, ... of the time of the program with unsorted data to calculate the same density.

In fact in the area above the thick line the maximum ratio is less than 1.1, which means that the program with sorted data needs only 10% more time to calculate the density than the program with unsorted data. The maximum ratio is lying in the upper left corner, which means the worst case happens if we have a small amount of datapoints and a big bandwidth. But the most interesting bandwidths for the datasets we investigated are found in the area between 0.01 and 1.0 (in the figure it is the range between $-2 = \log_{10}(0.01)$ and $0 = \log_{10}(1.0)$). If we reach the upper border, we are oversmoothing; if we reach the lower border, we are undersmoothing.

So, as a consequence, we will assume for all following programs that the data are sorted.

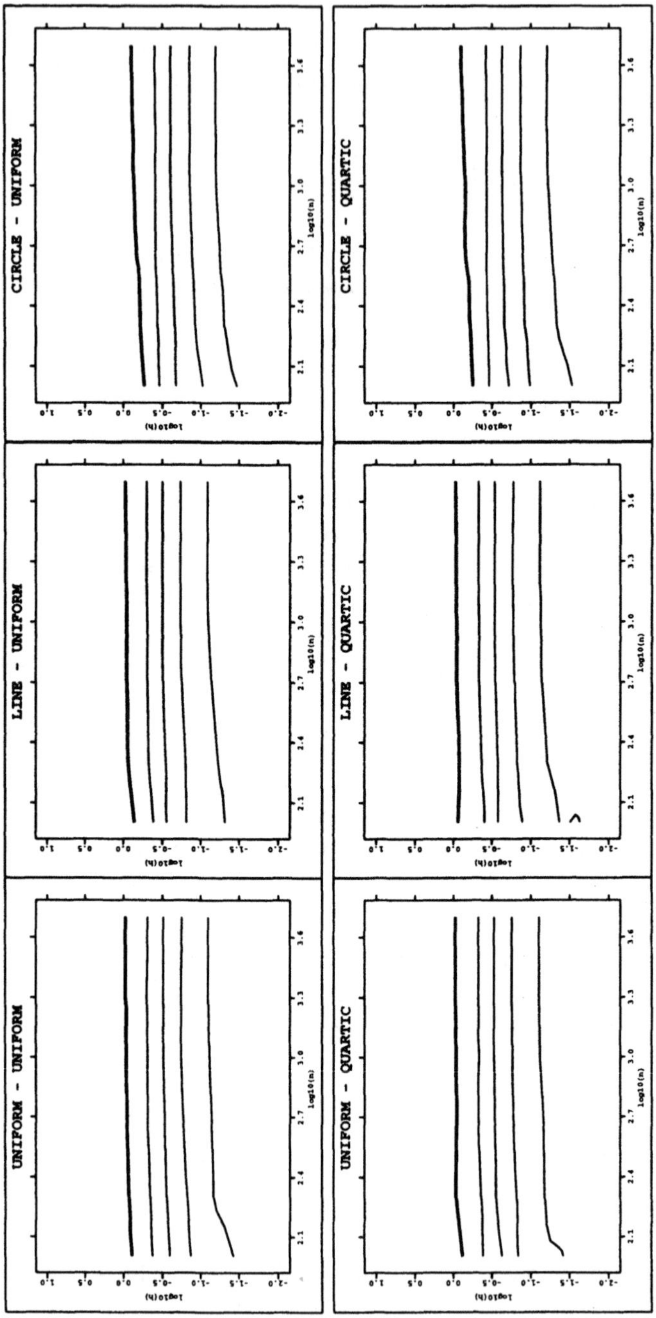

FIGURE 4.16. Relative computational time for density estimation of sorted and unsorted data

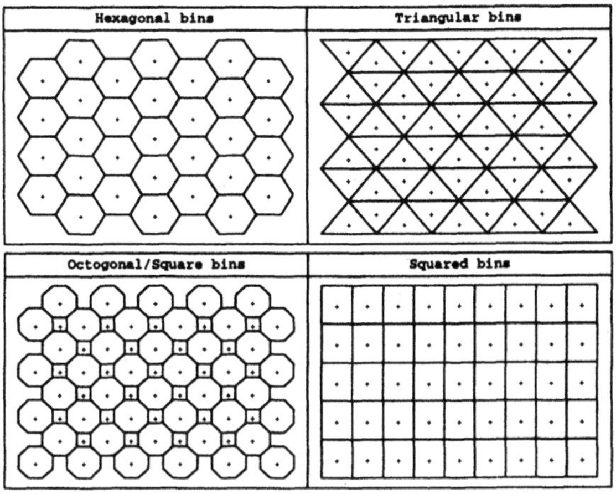

FIGURE 4.17. Possible tesselations of the plane

Binning

The main advantage of binning is that it reduces the computational costs of density (and regression) estimates by loosing only very little accuracy.

The binning first makes a "regular" tesselation of the real space $\mathbb{R}^p$, if p is the dimension of the data X. Every part of the tesselation, also called bin or cell, has to be connected and disjoint. Additionally the bins will exhaust the $\mathbb{R}^p$. One point, the bincenter is then taken to represent all the datapoints which are falling into such a bin. Usually the gravity center of each bin is taken. The set of bincenters S have the property, that for every bincenter the direct neighbours have the same distance. In the one dimensional case such tesselations are the equally spaced intervals.

In the one dimensional case for density estimation we profit mainly from a decrease of datapoints, and we even get one further advantage: the distance between bincenters can be expressed as multiples of the binwidth $\delta = \min_{p_1,p_2 \in S; p_1 \neq p_2} \mid p_2 - p_1 \mid$. So we have to calculate the kernel at the points $i\,\delta$ with $i = 0, 1, 2, \ldots$.

The extension of the tesselations in the two or multidimensional space raises some questions. Usually quadratic (or rectangular) bins are used in higher dimensions. But there are also other possibilities available as can be seen in Figure 4.17.

There are even possibilities with "nonregular" bins, as we see in the left picture (octogonal/square bins) of Figure 4.17.

A hexagonal tesselation seems to be more appropriate in our case, because we do not have to store as many zeros as when using a tesselation of squares. But with a hexagonal tesselation we have one problem: the binning algorithm and the density estimates become more complicated (see Appendix C). In this case it is possible to bin with the following precept:

The datapoint falls into the bin which has the nearest (weighted) bincenter.

To make life easier, we will restrict ourselves to squared bins. Another problem that arises in higher dimensions ($p > 3$) is that there are only two tesselations which have just one kind of symmetric polyhedron bin (like the archimedean bodies in $\mathbb{R}^3$). These are tesselations built up from hypercubes and hypertetrahedra.

In the multidimensional case the advantage of reducing the sample size is lost. If we take a small grid with 50 bins in every variable, we will get $50 \times 50 = 2500$ bins for a squared grid in the two dimensional case. If we have a dataset with 1000 observations we will get, on an average, 20 observations per bin in the one dimensional case. In the two dimensional case we will get 0.4 observations per bin.

To get an impression how fast the binning method works in two dimensions for different bandwidths and different sample sizes, I wrote a program, which calculates the densities directly, and another which uses binning (2500 bins). A comparison of computing times is shown in Figure 4.18. The thick line here always indicates that the ratio of computational time of the binned version and the unbinned version is 1. The thin lines above the thick line indicates a ratio of 0.75, 0.5, ..., which means that the binned program needs only 75%, 50%, ... of the time to calculate the same density. Under the thick line we see the ratio of 1.25, 1.5, ...

This has to be compared with Figure 3a in the appendix of Fan & Marron (1994), which for the one dimensional case shows a speed improvement of factor 10 and more for binned density estimation over direct density estimation. That would mean that the ratio becomes less than 0.1. Nevertheless in the interesting area between -1.5 and 0.0 (the common logarithm of the bandwidth) the ratio is on an average size of 0.75, and we gain some advantage using binning.

In Figure 4.19 the accuracy of the binned version against the unbinned version for the average Friedman-Tukey index is shown (our main interest are the projection pursuit indices!). The relative error ε_{FT} is calculated by:

$$\varepsilon_{FT} = \frac{|\,\overline{FT}_{binned} - \overline{FT}_{unbinned}\,|}{|\,\overline{FT}_{unbinned}\,|}$$

An error of 1% means $\varepsilon_{FT} = 0.01$. As mentioned above we have taken the

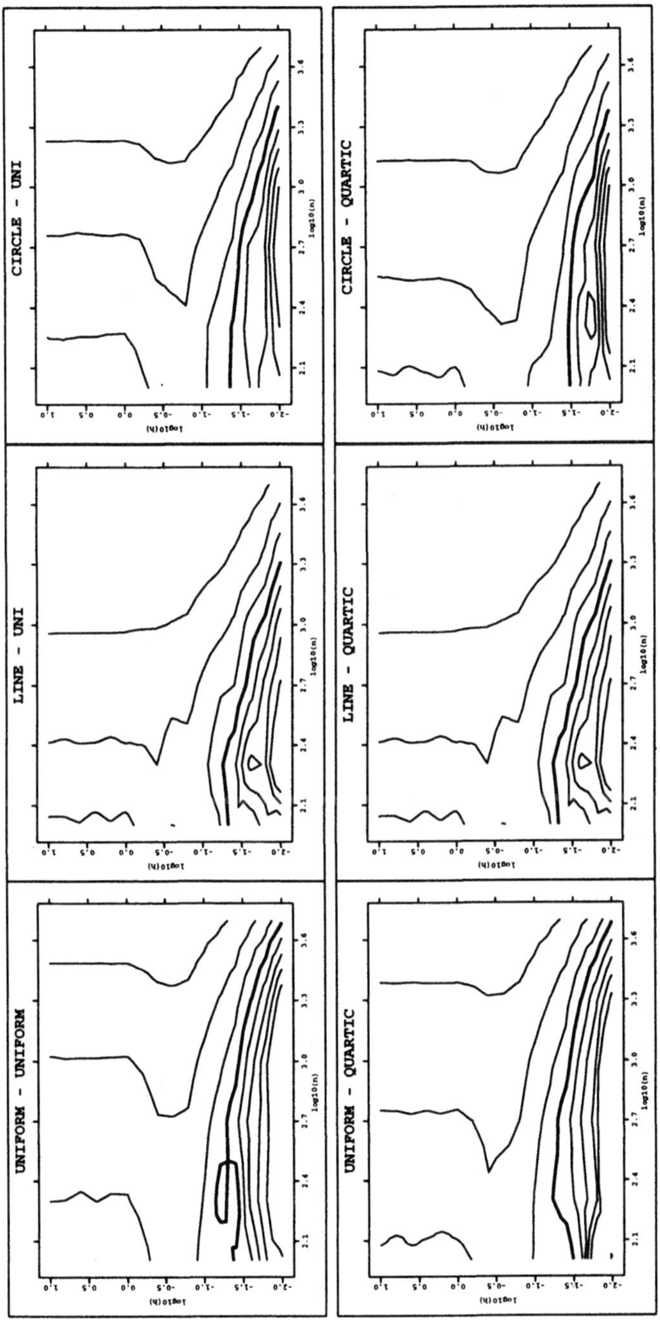

FIGURE 4.18. Relative computational time for the Friedman-Tukey index with binned and unbinned data.

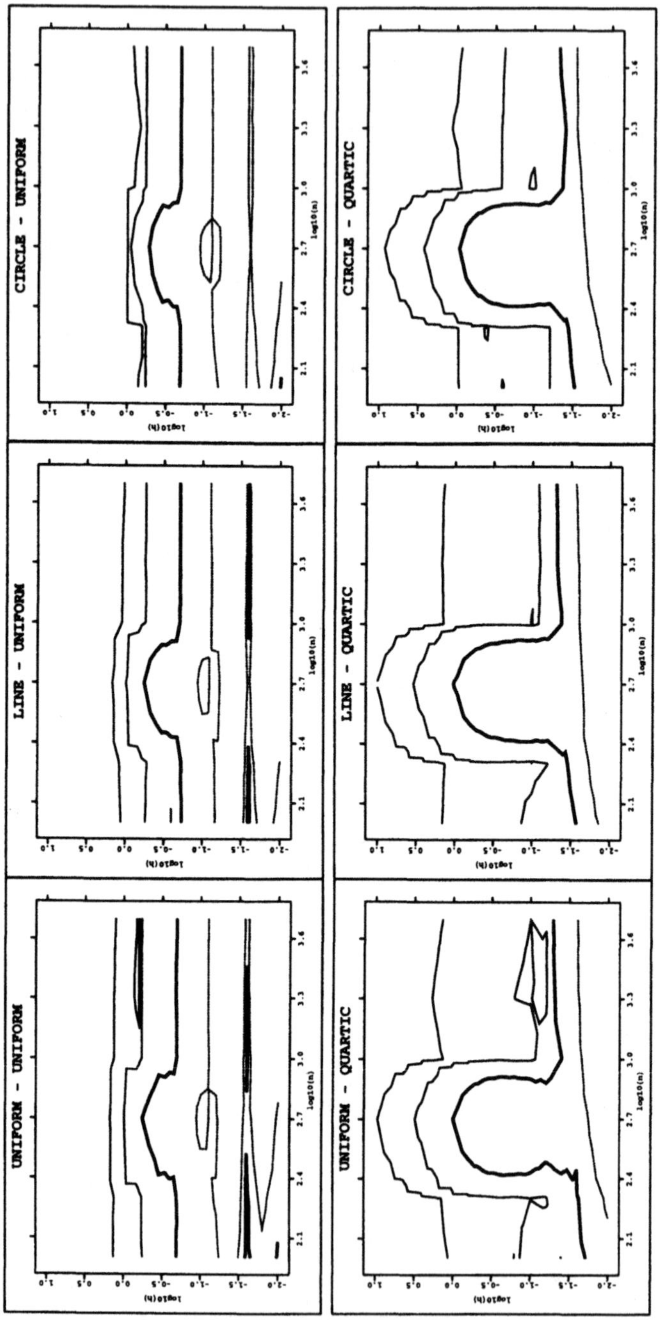

FIGURE 4.19. Relative error for the Friedman-Tukey index with binned and unbinned data.

average over a lot of Friedman-Tukey indices. It was necessary because the computational time for the density for 100 datapoints was less than the minimal measuretime (1/18 sec.). So we did a lot of loops and divided by the number of loops afterwards. From the top to the bottom we see the 0.01%-, 0.1%-, 1%- and 5%-line. The 1% is marked with a thick line, and 5% is the dotted line.

The replacment of the (continuous) kernel by a discret step function on a square will cause the value of the index function to vary if we rotate the data. As can be seen in Figure 4.20, the value of the density will change and so will the value of the index function, if a datapoint moves from one bin into another.

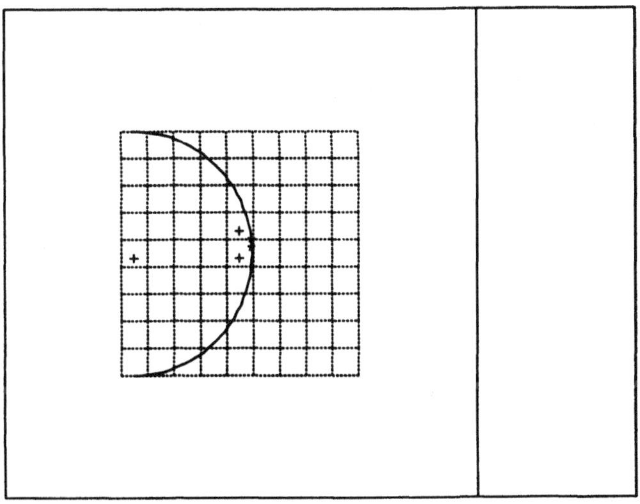

FIGURE 4.20. Rotation moves the star marked with point into another bin.

Assumed the datapoint marked with a star is rotated a little on the circle around the left cross to the second point marked with a star, then the bin-centers of these two points are represented by the two right points marked with a cross. The Friedman-Tukey index for this dataset would change from $2(K_2(0) + K_2(4))$ to $2(K_2(0) + K_2(\sqrt{17}))$, which will be different. As long as the kernels are continuous, this will have only little effect, but if we use the uniform kernel the effect will be drastic.

We can use linear binning which means replacing the kernel function by a stepwise linear function instead of a stepwise constant function. This would be done so that we distribute the datapoints on two (in the univariate case) or four bins (in the bivariate). From a dataset of 600 observations a colleague of mine could produce a binned dataset of 900 observations, so we have to stick to constant binning.

Reducing to integer operations

We know that estimating the density with a kernel is a task of summing up kernels at different points. So we will have a look at the summation of two or more different kernel values.

We will assume that the values r_i with $i = 1, 2, \ldots$ at which the kernels will be evaluated are less than 1. Moreover we can separate the bivariate kernels in an index function $I(r < 1)$ and the kernel function $K_{kern}(r)$ so that:

$$K_2(r) = I(r < 1)K_{kern}(r)$$

with $kern \in$ {uni, tri, epa, qua, triw, cos, log1, log2}. For all kernels we can find the following formulae:

Uniform

$$\begin{aligned} K_2(r_1) + K_2(r_2) &= K_{\text{uni}}(r_1) + K_{\text{uni}}(r_2) \\ &= \frac{1}{\pi} + \frac{1}{\pi} \\ &= 2K_{\text{uni}}(0) \\ \sum_{i=1}^{m} K_2(r_i) &= mK_{\text{uni}}(0) \end{aligned}$$

Triangle

$$\begin{aligned} K_2(r_1) + K_2(r_2) &= K_{\text{tri}}(r_1) + K_{\text{tri}}(r_2) \\ &= \frac{3}{\pi}(1 - r_1) + \frac{3}{\pi}(1 - r_2) \\ &= \frac{3}{\pi}(2 - r_1 - r_2) \\ &= \frac{3}{\pi} + \frac{3}{\pi}(1 - (r_1 + r_2)) \\ &= \frac{3}{\pi} + K_{\text{tri}}(r_1 + r_2) \\ \sum_{i=1}^{m} K_2(r_i) &= \frac{3(m-1)}{\pi} + K_{\text{tri}}\left(\sum_{i=1}^{m} r_i\right) \end{aligned}$$

Epanechnikov

$$K_2(r_1) + K_2(r_2) = K_{\text{epa}}(r_1) + K_{\text{epa}}(r_2)$$

$$
\begin{aligned}
&= \frac{2}{\pi}(1-r_1^2)+\frac{2}{\pi}(1-r_2^2) \\
&= \frac{2}{\pi}(2-r_1^2-r_2^2) \\
\sum_{i=1}^{m} K_2(r_i) &= \frac{2}{\pi}\left(m-\sum_{i=1}^{m} r_i^2\right)
\end{aligned}
$$

Quartic

$$
\begin{aligned}
K_2(r_1)+K_2(r_2) &= K_{\text{qua}}(r_1)+K_{\text{qua}}(r_2) \\
&= \frac{3}{\pi}(1-r_1^2)^2+\frac{3}{\pi}(1-r_2^2)^2 \\
&= \frac{3}{\pi}(2-2(r_1^2+r_2^2)+(r_1^4+r_2^4)) \\
\sum_{i=1}^{m} K_2(r_i) &= \frac{3}{\pi}\left(m-2\left(\sum_{i=1}^{m} r_i^2\right)+\left(\sum_{i=1}^{m} r_i^4\right)\right)
\end{aligned}
$$

Triweight

$$
\begin{aligned}
K_2(r_1)+K_2(r_2) &= K_{\text{triw}}(r_1)+K_{\text{triw}}(r_2) \\
&= \frac{4}{\pi}(1-r_1^2)^3+\frac{4}{\pi}(1-r_2^2)^3 \\
&= \frac{4}{\pi}(2-6(r_1^2+r_2^2)+6(r_1^4+r_2^4)-(r_1^8+r_2^8)) \\
\sum_{i=1}^{m} K_2(r_i) &= \frac{4}{\pi}\left(m-6\left(\sum_{i=1}^{m} r_i^2\right)+6\left(\sum_{i=1}^{m} r_i^4\right)-\left(\sum_{i=1}^{m} r_i^8\right)\right)
\end{aligned}
$$

Cosine

$$
\begin{aligned}
K_2(r_1)+K_2(r_2) &= K_{\cos}(r_1)+K_{\cos}(r_2) \\
&= \frac{1}{4(1-2/\pi)}\cos\left(\frac{\pi r_1}{2}\right)+\frac{1}{4(1-2/\pi)}\cos\left(\frac{\pi r_2}{2}\right) \\
&= \frac{2}{4(1-2/\pi)}\cos\left(\frac{\pi}{4}(r_1+r_2)\right)\cos\left(\frac{\pi}{4}(r_1-r_2)\right) \\
&= 8\left(1-\frac{2}{\pi}\right)K_{\cos}\left(\frac{r_1+r_2}{2}\right)K_{\cos}\left(\frac{r_1-r_2}{2}\right) \\
\sum_{i=1}^{m} K_2(r_i) &= ?
\end{aligned}
$$

The formula for transforming a sum of two cosines is:

$$
\cos(\alpha)+\cos(\beta)=2\cos\left(\frac{\alpha+\beta}{2}\right)\cos\left(\frac{\alpha-\beta}{2}\right).
$$

There is a formula to transform a product of 3 cosines into a sum of 4 cosines:

$$\begin{aligned} 4\cos(\alpha)\cos(\beta)\cos(\gamma) &= \cos(\alpha+\beta-\gamma)+\cos(\beta+\gamma-\alpha) \\ &\quad +\cos(\alpha+\gamma-\beta)+\cos(\alpha+\beta+\gamma). \end{aligned}$$

We have investigated these formulae and found that they come from an ably treatment of exponentials ($\cos(x) = 0.5(\exp(ix) + \exp(-ix))$). The deeper reason for having no summation formula is that there is no formula for a sum of exponentials. This was my main reason to exclude the gaussian kernel.

In the following step we take into account that we are interested in calculating the kernel values at binned datapoints $(b_{i,1}\delta/h, b_{i,2}\delta/h)$. We replace δ/h by τ and take the radii $r_i = \tau\sqrt{b_{i,1}^2 + b_{i,2}^2}$. Now we first exclude the uniform kernel because there is nothing to do; second the cosine kernel because no summation is possible; third the triangle and cosine because we come to a summation of squareroots, which can not be simplified:

Epanechnikov

$$\begin{aligned} \sum_{i=1}^{m} K_2(r_i) &= \frac{2}{\pi}\left(m - \tau^2 \sum_{i=1}^{m}(b_{i,1}^2 + b_{i,2}^2)\right) \\ &= \frac{2m}{\pi} - \frac{2\tau^2}{\pi}\left(\sum_{i=1}^{m}(b_{i,1}^2 + b_{i,2}^2)\right) \end{aligned}$$

Quartic

$$\begin{aligned} \sum_{i=1}^{m} K_2(r_i) &= \frac{3}{\pi}\left(m - 2\tau^2\left(\sum_{i=1}^{m}(b_{i,1}^2 + b_{i,2}^2)\right)\right. \\ &\quad \left. +\tau^4\left(\sum_{i=1}^{m}(b_{i,1}^2 + b_{i,2}^2)^2\right)\right) \\ &= \frac{3m}{\pi} - \frac{6\tau^2}{\pi}\left(\sum_{i=1}^{m}(b_{i,1}^2 + b_{i,2}^2)\right) \\ &\quad +\frac{3\tau^4}{\pi}\left(\sum_{i=1}^{m}(b_{i,1}^2 + b_{i,2}^2)^2\right) \end{aligned}$$

Triweight

$$\begin{aligned}\sum_{i=1}^{m} K_2(r_i) &= \frac{4}{\pi}\left(m - 6\tau^2\left(\sum_{i=1}^{m}(b_{i,1}^2 + b_{i,2}^2)\right)\right.\\ &\quad +6\tau^4\left(\sum_{i=1}^{m}(b_{i,1}^2 + b_{i,2}^2)^2\right)\\ &\quad \left.-\tau^8\left(\sum_{i=1}^{m}(b_{i,1}^2 + b_{i,2}^2)^4\right)\right)\\ &= \frac{4m}{\pi} - \frac{24\tau^2}{\pi}\left(\sum_{i=1}^{m}(b_{i,1}^2 + b_{i,2}^2)\right)\\ &\quad +\frac{24\tau^4}{\pi}\left(\sum_{i=1}^{m}(b_{i,1}^2 + b_{i,2}^2)^2\right)\\ &\quad -\frac{4\tau^8}{\pi}\left(\sum_{i=1}^{m}(b_{i,1}^2 + b_{i,2}^2)^4\right)\end{aligned}$$

For the epanechnikov, the quartic and the triweight kernel the summation reduces to a polynomial in τ^2. The coefficient of the polynomial is a product of a real number and an integer.

Before we come to the result for the computational time, we will mention one problem that arises. In the worst case, all datapoints are lying on a circle except one which is lying in the middle of the circle. So we obtain an upper bound of $(n-1)\left(\frac{h}{\delta}\right)^{2m}$ for the coefficients of τ^{2m}. When a dataset is selected, we can give an upper limit for the value of $\frac{h}{\delta}$, which is equal to the maximum of the range of each variable divided by the binwidth. So if we select a binwidth of 0.02 and know that the data are lying in $[0,1]^2$, it follows that $\frac{h}{\delta} < 50$.

The maximum figure which can be represented by a l-bit integer figure (8 bit = 1 byte) is 2^l. Thus to represent the upper border we need

$$\log_2\left((n-1)\left(\frac{h}{\delta}\right)^{2m}\right) = \log_2(n-1) + 2m\log_2\left(\frac{h}{\delta}\right)$$

bits. Unfortunately in a computer an integer will normally have 16 bits and/or 32 bits.

Figure 4.21 shows the sample size in the x-axis, the number of bits which are needed to represent the upper bound is shown in the y-axis. The dotted lines are for τ^2, the solid line for τ^4 for 625 ($= 25 \times 25$), 2.500, 10.000 and 40.000 bins. We see if we have more than 600 datapoints and 2.500 bins our summation will probably cause an overflow for the coefficients of τ^4.

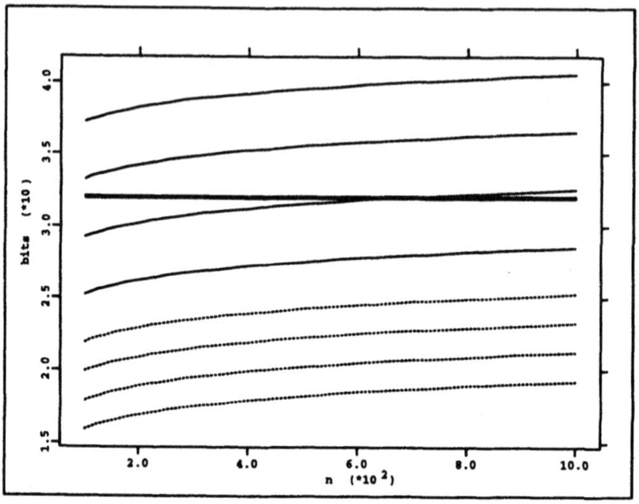

FIGURE 4.21. The maximum sample size for the integer coefficients for τ^{2m}

In pictures of Figure 4.22 we compare the relative computing time of the density estimation using these integer coefficients against the time needed by the program with the binned version. Again the thick line indicates a ratio of 1.0. The thin solid lines are indicating ratios of 0.8, 0.6, ..., which means that the integer-coefficient version is faster than the binned version. The dotted line indicates a ratio of 1.2.

As we see we gain in fact some computational time, especially if we have only few observations ($n < 500$).

But we also have to look at the relative error for the estimated Friedman-Tukey index, which is shown in Figure 4.23. The results for the uniform kernel are not shown because they are the same as in the binned version. As written above we run into trouble if we calculate the coefficients of τ^4. All pictures for the quartic kernel show a big relative error ε_{FT} if we have more than 600 datapoints.

In Figure 4.23 the thin solid line indicates a relative error of ..., 0.01%-, 0.1%. The lines can not be distinguished. The thick line indicates an error of 1%. The dotted lines indicate an error of 10% and 100%.

The result is disappointing both for computational time and accuracy. The reason for this disappointing result is that we were forced to use a long integer. The computation of calculations with long integers (4-byte) is faster than that with float-values, but slower than those using 2-byte integers. If we

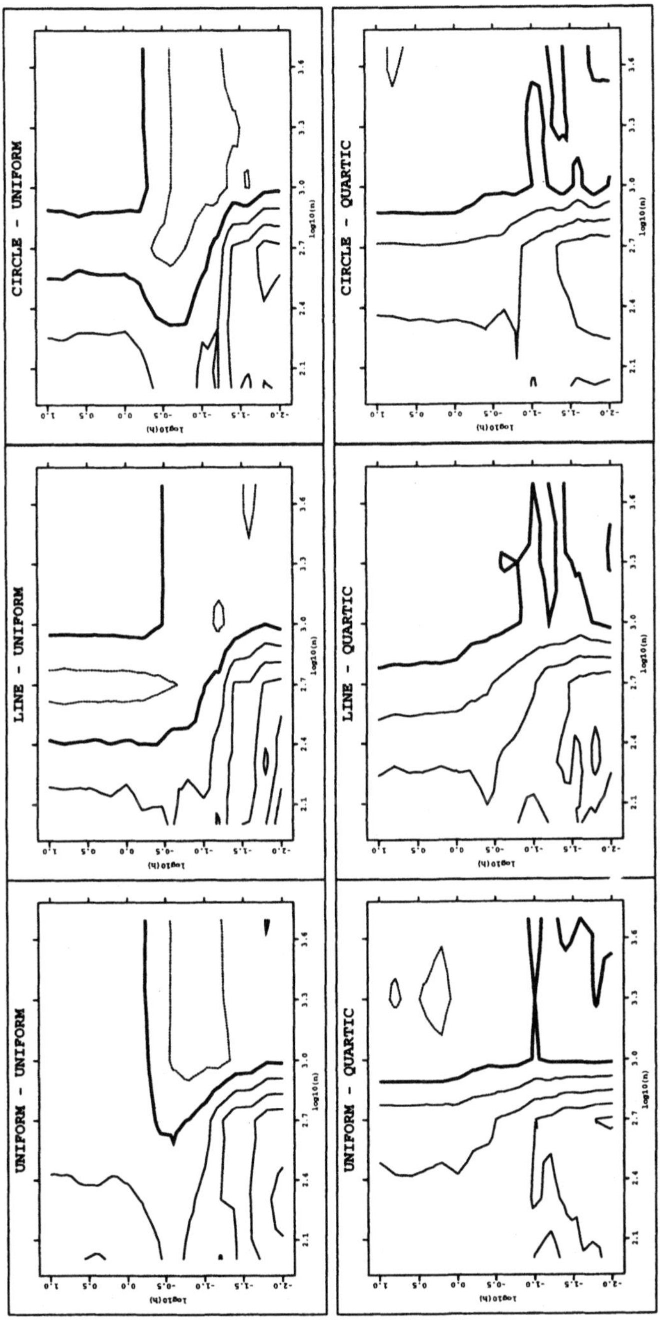

FIGURE 4.22. Relative computational time for the Friedman-Tukey index with binned data using integer-operations and float-operations.

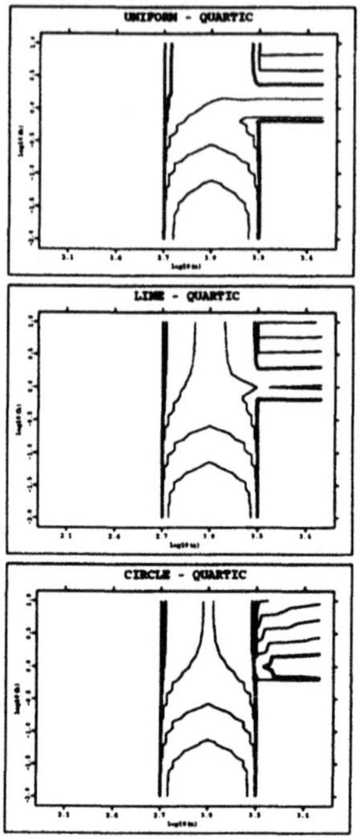

FIGURE 4.23. Relative error for the Friedman-Tukey index using integer-operations for the quartic kernel.

compare the inner loops for the density estimation with the quartic kernel, we would either have to carry out four float-operations or eleven long-integer operations.

Here is more potential for further optimization; for the Friedman-Tukey index and the quartic kernel, we get:

$$\begin{aligned}\hat{I}_{FT}(\alpha,\beta) &= \frac{1}{nh^2}\sum_{i=1}^{n}\left(\frac{3}{\pi}\left(n_i - 2\left(\sum_{j=1}^{n_i} r_{i,j}^2\right) + \left(\sum_{j=1}^{n_i} r_{i,j}^4\right)\right)\right)\\ &= \frac{3}{nh^2\pi}\left(\left(\sum_{i=1}^{n} n_i\right) - 2\left(\sum_{j=1}^{n_i} r_{i,j}^2\right) + \left(\sum_{j=1}^{n_i} r_{i,j}^4\right)\right),\end{aligned}$$

where n_i is the number of datapoints which have the distance $r_{i,j} < h$ to the datapoint X_i. We did not optimize any further because we get even with the quartic kernel overflows for the density estimation, and summing up more distances $r_{i,j}^4$ would diminish the sample size where we could obtain correct results.

It seems that replacing float-operations by integer-operations is not very successful, except if the sample size is small ($n < 250$). First we have to ensure that we do not get a (long-) integer overflow. That means the sample size is restricted even with a moderate number of bins. The second point is that we have to restrict ourselves to some kernels. The infinite support and the lack of a summation formula expelled the gaussian kernel. The difficulties with the summation of different kernel values force us to expel the triangle kernel as well. For the `XGobi`-program, for example, the triweight kernel is used. The reason for this is that the partial derivatives of the projection vectors which are used to speed up the maximization process, should be smooth enough. But we have to expel this kernel too, because with the binwidth used (0.02) the sample size has to be less than 2, otherwise a long-integer overflow could happen ($50^8 \gg 2^{32}$).

This result is not too bad for the quartic and epanechnikov kernel, although we are restricted as to the sample size. Even for small sample sizes the standard techniques like sorting and binning give some improvements in terms of computational time.

For small sample sizes ($n < 500$) the binning time also plays an important role. We replaced the binning algorithm by a simple truncation and we lost the advantage that more than one datapoint falls into one generated bin. But for these sample sizes we do not expect that we have a lot of bins with more than one observation. In Table 4.8 we give the number of nonempty bins (on the average) for all the datasets. Especially when we have only few observations we can see a further improvement of computational speed.

Observations	Uniform	Normal	Line	Two mass	Circle
100	98.0	93.3	94.6	94.6	96.8
200	191.9	172.6	179.0	178.1	186.2
500	451.5	353.7	382.4	375.5	416.2
1000	820.6	552.4	612.4	605.2	705.0
2000	1367.5	774.0	859.5	866.0	1061.0
5000	2138.0	1072.0	1117.0	1211.0	1457.0

TABLE 4.8. Number of nonempty bins (total number of bins: 2500)

Polynomial based indices

Compared to the kernel based indices the polynomial based indices are quite fast.

The orthogonal polynomials $f_n(x)$ applied in the approximation of the indices are defined by

$$\int_a^b \omega(x) f_n(x) f_m(x) dx = \begin{cases} 0 & n \neq m \\ h_n & n = m \end{cases} \quad \text{with } n, m \geq 0.$$

See Table 4.9 for a, b, and h_n.

	a	b	$\omega(x)$	h_n
Legendre polynomials P_n	-1	1	1	$\frac{2}{2n+1}$
Hermite polynomials H_n	$-\infty$	∞	e^{-x^2}	$\sqrt{\pi}2^n n!$
Natural-Hermite polynomials H_{e_n}	$-\infty$	∞	$e^{-x^2/2}$	$\sqrt{2\pi}n!$

TABLE 4.9. Creating values a, b, and h_n for orthonormal polynomials.

Legendre polynomial	$P_j(x) = \frac{(2j+1)P_{j-1}(x) - jP_{j-2}(x)}{j+1}$
Hermite polynomial	$H_j(x) = 2xH_{j-1}(x) - 2(j-1)H_{j-2}(x)$
Natural-Hermite polynomial	$H_{e_j}(x) = xH_{e_{j-1}}(x) - (j-1)H_{e_{j-2}}(x)$

TABLE 4.10. Recursive relationship for the polynomials used in the polynomial based indices.

The underlying recursive relationship of the orthogonal polynomials

$$f_n(x) = (a_n x + b_n) f_{n-1}(x) + c_n f_{n-2}(x)$$

allows a fast computation of individual values $f_n(Y_i)$. For the polynomials used in the indices the recursive relationship is given in Table 4.10. The constants a_n, b_n and c_n can be calculated by the appropriated scalarproduct given by the formula 4.2.5

$$< f_n(x), f_m(x) >= \int_a^b \omega(x) f_n(x) f_m(x) dx.$$

A detailed example can be found in the section 4.4.

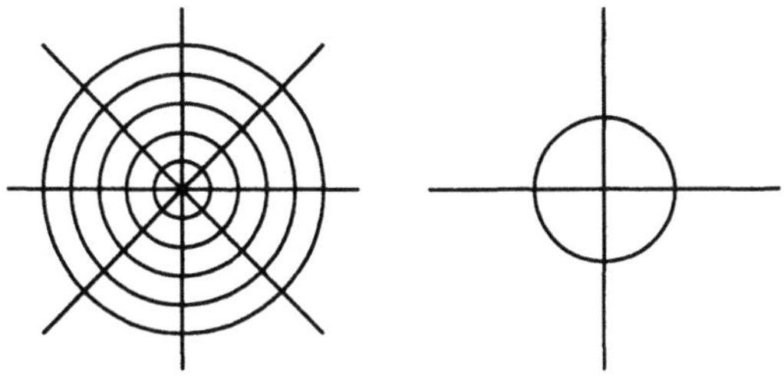

FIGURE 4.24. Left picture: Binning of the $\mathbb{R}^2$ plane as in Posse (1995a) so that all 48 bins have approximately the same weight under the standard normal distribution. The radius of the outer circle is equal to $R = \sqrt{2\log(6)}$, the radii of the inner circles are equal to $R/5, ..., 4R/5$. Right picture: Binning of $\mathbb{R}^2$ plane for the XploRe-macro PPINTER used to construct a χ^2-test.

4.2.6 *Limits*

Of course we can run in serious problems in EPP. We always have to answer the question if the "best" projection we have found is a real structure or just caused by random effects. The projections we get in the next section are very clear and we do not need to test them. But in general it is good idea to test the resulting projection for normality. In both papers of Posse (1995a, 1995b) about a new index, tests are used to indicate that the structure found is real.

Tests on multivariate normality are well known. In our case we have to exclude tests which use the first or second moments since our data are sphered.

The simplest test is a χ^2-test: we tesselate the plane as in Figure 4.24 and examine the the number of observations in each bin and compare it to the expected number of observations. A practical assumption of the χ^2-test is that we have at least five observations per bin. In our implementation we use only eight bins and it holds:

$$8/n\sum_{i=1}^{8}(\text{obs. in bin } i - n/8)^2 \sim \chi^2_7.$$

Other tests for a multivariate normal distribution rely on the third and fourth moments, e.g.

1. the (asymptotic) Mardia test on skewness (Mardia 1980)

$$\frac{1}{6n}\sum_{i,j=1}^{n}\left(x_{i,1}x_{j,1}+x_{i,2}x_{j,2}\right)^3 \sim \chi^2_4,$$

2. the (asymptotic) Mardia test on kurtosis

$$\frac{1}{n}\sum_{i=1}^{n}\left(x_{i,1}^2+x_{i,2}^2\right)^2 \sim N(8,64/n)$$

3. and the (asymptotic) tests on skewness and kurtosis (Lütkepohl 1990)

$$\lambda_1 = \frac{1}{6n}\left(\left(\sum_{i=1}^{n} x_{i,1}^3\right)^2 + \left(\sum_{i=1}^{n} x_{i,2}^3\right)^2\right) \sim \chi^2_2$$

$$\lambda_2 = \frac{1}{24n}\left(\left(\sum_{i=1}^{n}(x_{i,1}^4-3)\right)^2 + \left(\sum_{i=1}^{n}(x_{i,2}^4-3)\right)^2\right) \sim \chi^2_2$$

$$\lambda_1+\lambda_2 \sim \chi^2_4$$

The test of Baringhaus & Henze (1989) rely on the comparison of the empirical characteristic function with the characteristic function of the standard normal distribution

$$n\int_{I\!R^2} \mid 1/n\sum_{j=1}^{n}\exp(i(t_1x_{i,1}+t_2x_{i,2})) - \exp(-0.5(t_1^2+t_2^2)) \mid^2 \phi(t_1,t_2)dt$$

which leads to a test statistic

$$\frac{1}{n}\sum_{j,k=1}^{n}\exp(-0.5((x_{i,1}-x_{j,1})^2+(x_{i,2}-x_{j,2})^2)-\sum_{j=1}^{n}\exp(-0.25(x_{i,1}^2+x_{i,2}^2))+n/3.$$

Although the asymptotic distribution is rather complicated, the values for the quantiles can be found in Theilen (1990) calculated via simulations.

In general each test on normality or nonnormality can be used to build an index for exploratory projection pursuit.

4.3 Application to the Swiss Banknote Dataset

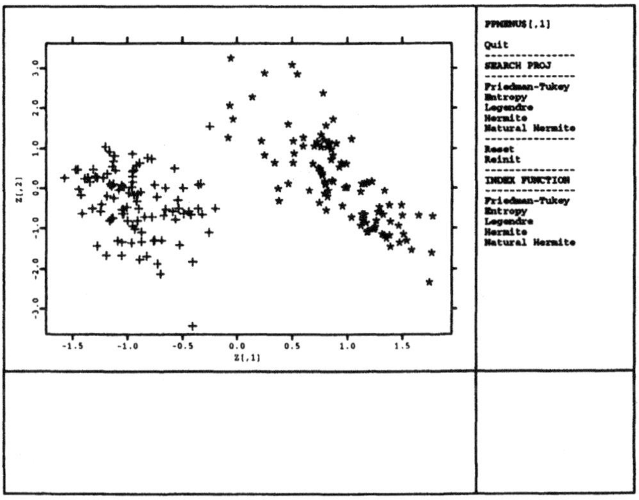

FIGURE 4.25. Basic screen of the PPEXPL macro. Swiss banknote data projected on the first two principal components. Crosses indicate genuine banknotes and stars indicate the forged bank notes

The dataset we use here is the Swiss banknote dataset. The dataset consists of 200 observations: 100 genuine and 100 forged banknotes. The six variables are measurements of the size of the banknotes (see Table A.4). Figure 4.25 shows a principal component plot of the first two principal components of the dataset. We can clearly distinguish the forged and genuine banknotes. If we use the Hermite index with order 7, we get the picture in Figure 4.26. This picture clearly shows three clusters: the upper one consists mostly of forged bank notes and only one genuine bank note (observation 70).

The other clusters contain forged or genuine banknotes. An easy interpretation would be that at least two gangs of forgers were falsifying Swiss banknotes. It seems that one of the genuine banknotes has been classified wrongly.

The projection and loading vectors from the exploratory projection pursuit step are

$$p = \begin{pmatrix} 0.896562 & 0.381411 \\ -0.087805 & 0.632784 \\ -0.248246 & 0.423262 \\ -0.349619 & 0.510308 \\ -0.044133 & 0.113526 \\ 0.051564 & 0.040710 \end{pmatrix} \quad l = \begin{pmatrix} -0.1643288 & 0.0958421 \\ -0.4423175 & 0.3462294 \\ 0.0023849 & 0.6045508 \\ 0.7167979 & -0.6467759 \\ 0.6894302 & -0.5717335 \\ 0.2225492 & -1.2337853 \end{pmatrix}.$$

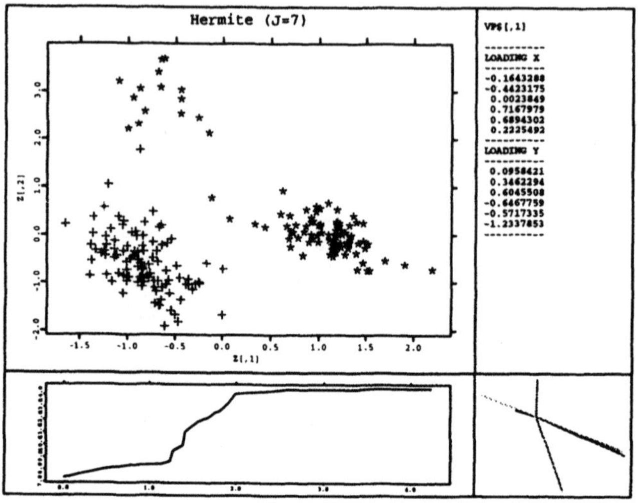

FIGURE 4.26. Projection with Hermite index (J=7). The right upper window shows the loadings with respect to X; the right lower window shows a graphical representation of the loadings. The left lower window shows how the index function has increased.

The projection vectors p could be interpreted as factor loadings with respect to the standardized variables Y; and the loading vectors l as factor loadings with respect to the original variables X. The first projection turns out to depend mainly on the first component of Y while the second reflects a mean of the second, third and fourth components of Y. An interpretation of $\alpha^T\Sigma^{-1/2}$, $\Sigma = Cov(X) = (\sigma_{ij})$, and $\alpha^T\Sigma^{-1/2}diag(\sigma_{ii})^{-1/2}$ as factor loadings with respect to the unsphered variables as in PCA based on covariance or correlation structure may be of interest.

Exploratory projection pursuit is a technique that allows us to use interactive graphics for interpretation. We can link the "best" projection with each of the variables (strips of data in lower plot in Figures 4.27 - 4.29). By simple linking we can visualize the relationship between the projection and the original variables. In the upper plot we see the projection found for the Swiss banknotes. The lower plot shows jittered dotplots of each of the six variables. The variables are rescaled on $[0, 1]$ via $xr_j = (x_j - \min_j(x_j))/(\max_j(x_j) - \min_j(x_j))$. The text window on the right shows the projection vector for X and Y found by exploratory projection pursuit.

We mark the three clusters, and they appear in univariate projections into the coordinate (variable) axes in order to assist interpretation based on differences of banknote sizes. Figure 4.27 shows clearly that all the points in the first cluster have low values in the variable 6 (inner box diagonal mea-

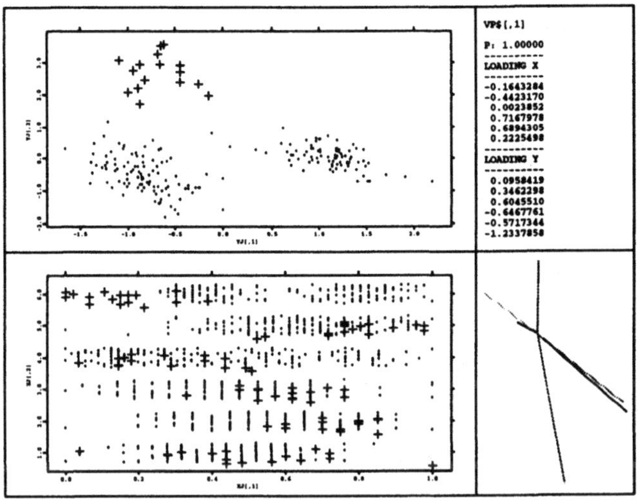

FIGURE 4.27. First cluster (forged banknotes) linked to original variables.

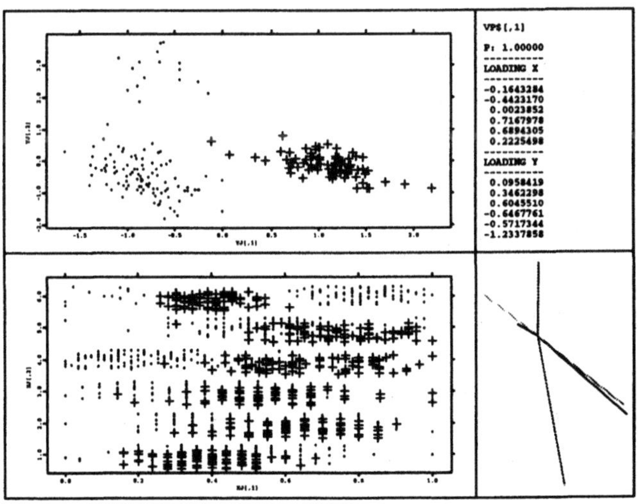

FIGURE 4.28. Second cluster (forged banknotes) linked to original variables.

surement of bank note). The second cluster seems to be linked to the high values of the sixth variable and the low variables of the fourth variable. The third cluster is linked to the low values of the sixth and the high values of

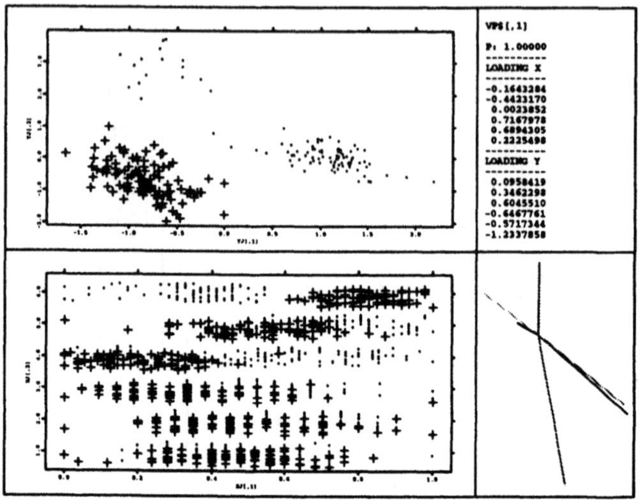

FIGURE 4.29. Third cluster (genuine banknotes) linked to original variables.

the fourth variable.

A possible next step would be to mark each variable to see how it behaves in the projection. However, we will not find anything interesting with this dataset.

4.4 Multivariate Exploratory Projection Pursuit

4.4.1 Why Multivariate Indices ?

A new set of indices will be proposed here. As univariate indices will mainly find univariate structures and bivariate indices will mainly find bivariate structures, a structure as in Figure 4.26 can not be found easily by a univariate index.

The same is, of course, true for three dimensional structures and for bi- or univariate indices. To find a d dimensional structure in a p dimensional dataset we need indices that are based on d dimensional projections.

It is clear that is not necessary to develop indices for dimensions bigger than ~ 6. If the dimension of the projection becomes too big we can not visualize the result easily, and furthermore the dataset has to be highdimensional. Otherwise we can not speak any longer of using "low dimensional" projections.

We can try to extend the kernel based indices, which is easily done by the multivariate kernels in Table 4.3. But here we run into serious problems known as the "curse of dimensionality". As Silverman (1986) pointed out the density estimation in the tails of the distribution needs a lot of observations. Silverman gave Table 4.11 to show how many observations are necessary to estimate $\phi(\mathbf{0})$ via kernel density estimation with a relative mean squared error less than 0.1. Note here that $\phi(\mathbf{0})$ is in the center of the distribution, and to estimate the tails of the distribution correctly a much larger sample would be needed.

Dimension	Required sample size
1	4
2	19
3	67
4	223
5	768
6	2790
7	10700
8	43700
9	187000
10	842000

TABLE 4.11. Required sample size to estimate $\phi(\mathbf{0})$ correctly

So, although the kernel based indices have the advantage of easy implementation and can compete in terms of computational speed with polynomial based indices, they are not useful for higher dimensional indices.

But if we look at the empirical form of the polynomial based indices we see easily that they have a complicated form. See for example the appendix of Polzehl & Klinke (1994) for the implementation of the empirical form and compare the polynomial based indices with kernel based.

Another disadvantage is that the computational costs for extending these indices increase polynomial. The number of coefficients that need to be calculated for an fixed order J is $\approx J^d/2$, where d is the dimension of the projection. We also have to avoid to estimate more parameters than data. In the case of $J = 10$ and $d = 3$ we have already 500 parameters to estimate. Thus we would need more than 500 observations to make such an estimate reliable.

As a consequence none of these indices seemed to be appropriate for an extension in higher dimensions.

4.4.2 A Class of Multivariate Indices

In general all of the indices are coding a d dimensional projection into a figure which describes the amount of structure. The amount of structure is described by a wide departure from the normal distribution.

The requirement of the rotation invariance of the indices can be expressed easily by using polar coordinates. Rotation invariance means that the distance between 0 and the datapoints (the radius) and the angels between the datapoints are constant under a rotation, so when we base the new indices on the radii the condition of rotation invariance is fulfilled. How would a normal distribution behave if we would look just at the radii ?

We know that when the random variables X_i are independent and standard normally distributed the distribution of

$$\sum_{i=1}^{d} X_i^2 \sim \chi_d^2$$

is a χ-square distribution with d degrees of freedom. The underlying density is given by

$$w_d(x) = \begin{cases} \frac{1}{2^{d/2}\Gamma(d/2)} x^{p/2-1} e^{-x/2} & x > 0 \\ 0 & x \leq 0 \end{cases}$$

New indices can be developed by finding functionals that minimize this univariate distribution just like the entropy index for the standard normal distribution. In fact we would like to develop the new indices analog to the polynomial based indices

$$I = \int_0^\infty (f(r) - w_d(r))^2 g(r) dr$$

with a weight function $g(r)$.

We can transform I into

$$\begin{aligned} I &= \int_0^\infty (f(r) - w_d(r))^2 g(r) dr \\ &= \int_0^\infty (f^2(r)g(r) - 2f(r)g(r)w_d(r) + w_d^2(r)g(r))dr \\ &= \int_0^\infty f(r)g(r)f(r)dr - 2\int_0^\infty w_d(r)g(r)f(r) + \int_0^\infty w_d^2(r)g(r)dr. \end{aligned}$$

If the integral of the last term exists, it is a constant and can be neglected, since we later will be interested in maximization of these indices. Thus we get

$$\begin{aligned} I &= \int_0^\infty f(r)g(r)\, f(r)dr - 2\int_0^\infty w_d(r)g(r)\, f(r)dr + C_1 \\ &= E(f(r)g(r)) - 2E(w_d(r)g(r)) + C_1 \end{aligned} \tag{4.5}$$

and we can replace the expectation by the appropriate sample mean

$$\tilde{I} = \frac{1}{n}\sum_{i=1}^{n} g(r_i)(f(r_i) - 2w_d(r_i)).$$

We now have to estimate the unknown density f by an estimator $\hat{f}$ to get a estimate $\hat{I}$ for I. Here we could use a kernel estimate, but we should keep in mind that the support of the density is limited by the left border 0. Another possibility would be boundary kernels as described in Scott (1992), but one of the aims we have is easy computation. Therefore it seems more appropriate to expand

$$f(r) = \sum_{i=0}^{\infty} a_i P_i(r)$$

with an orthonormal function system P_i, and we will get an estimator of order J

$$\hat{I} = \frac{1}{n}\sum_{i=1}^{n} g(r_i)\left(\sum_{j=0}^{J} \hat{a}_i P_j(r_i) - 2w_d(r_i)\right)$$

Estimation of the coefficients a_k can be done via

$$\begin{aligned} E(P_k(r)) &= \int_0^\infty f(r)P_k(r)dr \\ &= \int_0^\infty \left(\sum_{i=0}^{\infty} a_i P_i(r)P_k(r)\right) dr \\ &= \sum_{i=0}^{\infty} a_i \left(\int_0^\infty P_i(r)P_k(r)dr\right) \\ &= a_k \end{aligned}$$

and the estimate will be

$$\hat{a}_k = \frac{1}{n}\sum_{i=0}^{n} P_k(r_i).$$

The final estimate results in

$$\hat{I} = \frac{1}{n}\sum_{i=1}^{n} g(r_i)\left(\sum_{j=0}^{J} P_j(r_i)\left(\frac{1}{n}\sum_{k=0}^{n} P_j(r_k)\right) - 2w_d(r_i)\right). \tag{4.6}$$

The calculation of the function $P_k(r)$ can simply be derived from the fact that

$$\int_0^\infty P_i(r)P_j(r)dr = \delta_{ij},$$

thus if we define

$$P_i(r) = p_i(r)\sqrt{w_d(r)}$$

with $p_i(r)$ a polynomial in r we get

$$\int_0^\infty p_i(r)p_j(r)w_d(r)dr = \delta_{ij}.$$

We interpret this as a special scalarproduct

$$\langle f, g\rangle = \int_0^\infty f(r)g(r)w_d(r)dr$$

and define a fitting norm by

$$||f(r)|| = \sqrt{\langle f(r), f(r)\rangle}.$$

Now we can calculate the functions $p_i(r)$ via the orthogonalization method of Erhardt-Schmidt:

$$p_i(r) = \frac{r^i - \sum_{j=0}^{i-1}\langle r^i, p_j(r)\rangle p_j(r)}{||r^i - \sum_{j=0}^{i-1}\langle r^i, p_j(r)\rangle p_j(r)||}.$$

If follows that

$$p_n(r) = \frac{\displaystyle\sum_{i=0}^{n} \binom{n}{i} (-1)^{n-i} r^i \frac{\int_0^\infty t^n w_d(t)dt}{\int_0^\infty t^i w_d(t)dt}}{\sqrt{\int_0^\infty t^{n-1} w_d(t)dt \prod_{i=1}^{n}(2i)}}$$

with

$$\int_0^\infty r^n w_d(r)dr = \prod_{i=0}^{n-1}(p+2i).$$

A recursive relationship can be derived too. It follows from

$$P_i(r) = (a_i x + b_i)P_{i-1}(r) + c_i P_{i-2}(r)$$

via calculating of the scalarproducts

$$\begin{aligned}
1 &= \langle p_i(r), p_i(r)\rangle \\
&= \langle p_i(r), (a_i r + b_i)p_{i-1}(r) + c_i p_{i-2}(r)\rangle \\
&= a_i\langle p_i(r), rp_{i-1}(r)\rangle + b_i\langle p_i(r), p_{i-1}(r)\rangle + c_i\langle p_i(r), p_{i-2}(r)\rangle \\
&= a_i\langle p_i(r), rp_{i-1}(r)\rangle,
\end{aligned}$$

$$\begin{aligned}
0 &= \langle p_{i-1}(r), p_i(r)\rangle \\
&= a_i\langle p_{i-1}(r), rp_{i-1}(r)\rangle + b_i
\end{aligned}$$

$$\begin{aligned}
0 &= \langle p_{i-2}(r), p_i(r)\rangle \\
&= a_i\langle p_{i-2}(r), rp_{i-1}(r)\rangle + c_i.
\end{aligned}$$

We get the following formula for a_i, b_i and c_i

$$\begin{aligned}
a_i &= \frac{1}{\sqrt{(2i)(d+2i-2)}} \\
b_i &= -\frac{d+4i-4}{\sqrt{(2i)(d+2i-2)}}
\end{aligned}$$

$$c_i = -\sqrt{\frac{(i-1)(d+2i-4)}{i(d+2i-2)}}.$$

The **Mathematica** program in Appendix D.2.1 calculates quantities $p_i(r)$, a_i, b_i and c_i; the recursive formula can be evaluated to

$$p_i(r) = \frac{(r-(d+4i-4))p_{i-1}(r) - \sqrt{(2i-2)(d+2i-4)}p_{i-2}(r)}{\sqrt{(2i)(d+2i-2)}}.$$

4.4.3 Special Indices

Before developing special indices using different kind of weight functions, let us have a look at the behaviour. Assumed the random variable X_i has a density $f_i(x)$ we can assume that the density $f_i(x)$ is bounded in $I\!R$ and $\lim_{|x|\to\infty} f(x) = 0$ as we have finite data samples. We can calculate the density of X_i^2 by

$$\begin{aligned} E(X^2) &= \int_{I\!R} x^2 f(x)dx \\ &= \underbrace{\int_{-\infty}^{0} x^2 f(x)dx}_{x=-\sqrt{z}} + \underbrace{\int_{0}^{\infty} x^2 f(x)dx}_{x=\sqrt{z}} \\ &= \int_0^\infty z\frac{f(-\sqrt{z})}{2\sqrt{z}}dz + \int_0^\infty z\frac{f(\sqrt{z})}{2\sqrt{z}}dz \\ &= \int_0^\infty z\frac{f(\sqrt{z})+f(-\sqrt{z})}{2\sqrt{z}}dz \\ &= E(Z) \end{aligned}$$

and the density of the variable $Z_i = X_i^2$ will be

$$g_i(z) = \frac{f_i(\sqrt{z}) + f_i(-\sqrt{z})}{2\sqrt{z}}.$$

If we look at the univariate projections ($d = 1$) our index will be

$$I = \int_0^\infty \left(\frac{f_i(\sqrt{r}) + f_i(-\sqrt{r})}{2\sqrt{r}} - \frac{e^{-r/2}}{\sqrt{2\pi r}}\right)^2 g(r)dr$$

$$= \int_0^\infty \left(\frac{f_i(\sqrt{r}) + f_i(-\sqrt{r})}{2} - \frac{e^{-r/2}}{\sqrt{2\pi}} \right)^2 \frac{g(r)}{r} dr$$

Under the above assumption the squared term describes a finite difference, which goes to 0 if $r \to \infty$. The problem arises for $r \approx 0$. The term $1/r$ weights the differences near 0 much stronger than near $r = 1$. As a consequence the index will be attracted by a "central mass" distribution. We can avoid this by including an r in the weight function $g(r)$. In the multivariate case the development is much more difficult. Let us assume that the X_i will be independent and their density has a polynomial expansion

$$f_j(x) = \sum_{m=0}^{\infty} a_{m,j}(x - x_0)^m.$$

For simplicity we assume that $x_0 = 0$ and get the density $g_i(x)$ of X_i^2 as

$$g_i(x) = \sum_{m=0}^{\infty} c_{i,m} \frac{x^m}{\sqrt{x}}.$$

The density $h_d(x)$ of $R = X_1^2 + ... + X_d^2$ can be developed by

$$h_d(x) = x^{d/2-1} \sum_{m=0}^{\infty} p_{d,m} x^m.$$

For $d = 1$ we get

$$\begin{aligned} h_1(x) &= \frac{f_1(\sqrt{x}) + f_1(-\sqrt{x})}{2\sqrt{x}} \\ &= x^{-1/2} \sum_{m=0}^{\infty} \frac{a_{m,1}}{2} x^{m/2} (1 + (-1)^m) \\ &= x^{-1/2} \sum_{m=0}^{\infty} p_{1,m} x^m, \end{aligned}$$

with $p_{1,m} = c_{m,1} = a_{m,1}(1 + (-1)^m)/2$. Assume now that

$$h_{d-1}(x) = x^{d/2-1.5} \sum_{i=0}^{\infty} p_{i,d-1} x^i$$

and it follows

$$
\begin{aligned}
h_d(x) &= \int_0^x h_{d-1}(z) g_d(x-z)\,dz \\
&= \int_0^x \left(\sum_{i=0}^{\infty} p_{i,d-1} z^{i+d/2-1.5}\right)\left(\sum_{j=0}^{\infty} c_{j,d}(x-z)^{j-0.5}\right) dz \\
&= \sum_{i=0}^{\infty}\sum_{j=0}^{\infty} p_{i,d-1} c_{j,d} \int_0^x z^{i+d/2-1.5}(x-z)^{j-0.5} dz \\
&= \sum_{i=0}^{\infty}\sum_{j=0}^{\infty} p_{i,d-1} c_{j,1} x^{d/2-1} x^{i+j} \frac{\Gamma((d-1)/2+i)\Gamma(1/2+j)}{\Gamma(d/2+i+j)} \\
&= x^{d/2-1} \sum_{m=0}^{\infty} x^m \underbrace{\sum_{l=0}^{m} p_{l,d-1} c_{m-l,d} \frac{\Gamma((d-1)/2+l)\Gamma(1/2+m-l)}{\Gamma(d/2+m)}}_{p_{d,m}}.
\end{aligned}
$$

If we now compare the indices we get

$$
\begin{aligned}
I &= \int_0^{\infty} (f(r) - w_d(r))^2 g(r)\,dr \\
&= \int_0^{\infty} \left(r^{d/2-1}\tilde{f}(r) - \frac{1}{2^{d/2}\Gamma(d/2)} r^{d/2-1} e^{-r/2}\right)^2 g(r)\,dr \\
&= \int_0^{\infty} \left(\tilde{f}(r) - \frac{1}{2^{d/2}\Gamma(d/2)} e^{-r/2}\right)^2 r^{d-2} g(r)\,dr.
\end{aligned}
$$

We have some theoretical evidence that the index will be attracted by a "central" distribution for the univariate projection and attracted by skewed distributions for three- or higher dimensional projections. This behaviour can also be observed in practice. We can compensate this by choosing the weight function as shown in Table 4.12.

We will restrict ourselves to the three indices as the polynomial based indices, the

- χ^2_d-Legendre index
- χ^2_d-Hermite index
- χ^2_d-Natural-Hermite index

In the Legendre index of Friedman (1987) we have as weight function

d	$g(r)$
1	r
2	1
3	$1/r$
4	$1/r^2$
5	$1/r^3$
6	$1/r^4$

TABLE 4.12. Weight function to correct the index for a d dimensional projection

$$g(y) = \frac{1}{2\phi(y)}$$

and we choose the weight function for the χ_d^2-Legendre index as

$$g(r) = \frac{1}{2w_d(r)}.$$

The estimate (4.6) simplifies to

$$\begin{aligned} \tilde{I}_L &= \frac{1}{n}\sum_{i=1}^{n}\frac{1}{2w_d(r_i)}\left(\sum_{j=0}^{J}P_j(r_i)\left(\frac{1}{n}\sum_{k=0}^{n}P_j(r_k)\right) - 2w_d(r_i)\right) \\ &= -1 + \frac{1}{2n^2}\sum_{i=1}^{n}\sum_{j=0}^{J}\frac{P_j(r_i)}{w_d(r_i)}\left(\sum_{k=0}^{n}P_j(r_k)\right) \end{aligned}$$

and neglecting the term -1 leads to

$$\hat{I}_L = \frac{1}{2n^2}\sum_{i=1}^{n}\sum_{j=0}^{J}\frac{P_j(r_i)}{w_d(r_i)}\left(\sum_{k=0}^{n}P_j(r_k)\right).$$

The Hermite index of Hall (1989a) has as weight function

$$g(\mathbf{y}) = 1$$

and the weight function for the χ_d^2-Hermite index is

$$g(r) = 1,$$

which leads to a simplified form of (4.6)

$$\begin{aligned}\hat{I}_H &= \frac{1}{n}\sum_{i=1}^{n}\left(\sum_{j=0}^{J}P_j(r_i)\left(\frac{1}{n}\sum_{k=0}^{n}P_j(r_k)\right) - 2w_d(r_i)\right)\\ &= \frac{1}{n^2}\left(\sum_{j=1}^{J}\sum_{i,j=0}^{n}P_j(r_i)P_j(r_k)\right) - \frac{2}{n}\sum_{i=1}^{n}w_d(r_i),\end{aligned}$$

In the case of the Natural-Hermite index of Cook et al. (1993) we have as weight function

$$g(\mathbf{y}) = \phi(\mathbf{y})$$

and we take the weight function for the χ^2_d-Natural-Hermite index as

$$g(r) = w_d(r).$$

The estimate (4.6) simplifies to

$$\begin{aligned}\tilde{I}_N &= \frac{1}{n}\sum_{i=1}^{n}w_d(r_i)\left(\sum_{j=0}^{J}P_j(r_i)\left(\frac{1}{n}\sum_{k=0}^{n}P_j(r_k)\right) - 2w_d(r_i)\right)\\ &= \frac{1}{n^2}\left(\sum_{j=0}^{J}\sum_{i,k=1}^{n}w_d(r_i)P_j(r_i)P_j(r_k)\right) - \frac{2}{n}\sum_{i=1}^{n}w^2(r_i),\end{aligned}$$

Since the functional we have to calculate is complicated, we can improve the speed of the computation by binning. That means we replace

$$\bar{r}_i = \mathrm{int}(r_i/\delta),$$

with $\mathrm{int}(x)$ giving the largest integer smaller than x and δ being a binwidth. We can tabulate the functions $P_j(i\delta)$ and $w_f(i\delta)$.

4.4.4 Application to the Swiss Banknote Data

We apply the technique again to the Swiss banknote dataset. We know from the previous EPP that we can find at least 3 clusters in the data.

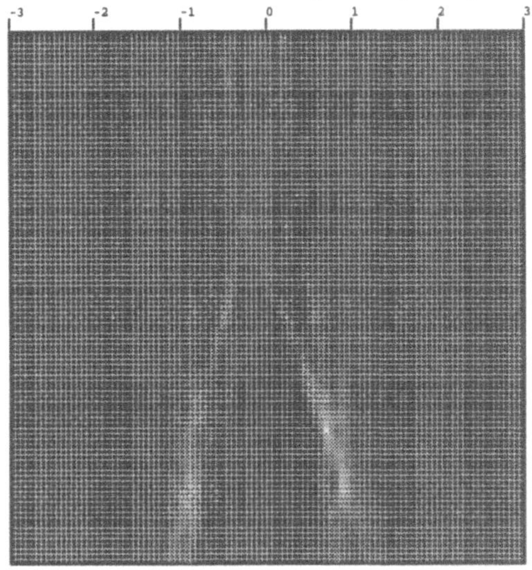

FIGURE 4.30. Projection based on the first (two) principal components. We can clearly distinguish two clusters. The chosen index is χ^2-Natural-Hermite index with order 15.

To visualize the results we use a modified parallel coordinate plot. We know that the most uninteresting density in the projection is the standard normal density, and we are able to say how the line densities will look like if we have a standard normal gaussian density. We compute the line densities and then subtract the normal densities. So the dark areas are those where the projected density is larger than the normal density, and the light areas are those where the projected density is smaller than the normal density.

We used a simulated annealing algorithm for finding the best projection. The program examines at least 1 million projections if does not stop before. Some pictures run up to 6 million projections. This is far more than the number of projections given in Table 4.1.

On most of the Figures 4.30-4.35 we are able to see two clusters. The left picture of Figure 4.30 and the right picture of Figure 4.34 might reveal three clusters.

The fact that in general we cannot find at least three clusters is due to the basic problem of these indices.

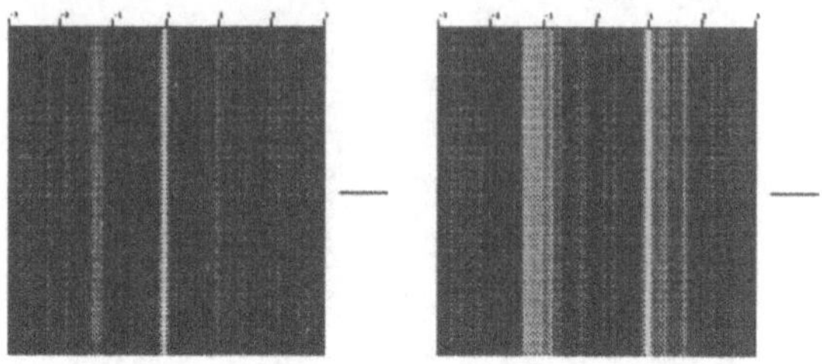

FIGURE 4.31. Result of optimizing the χ^2-Natural-Hermite index with order 10 (left picture) and χ^2-Hermite index with order 10 (right picture) for a univariate projection.

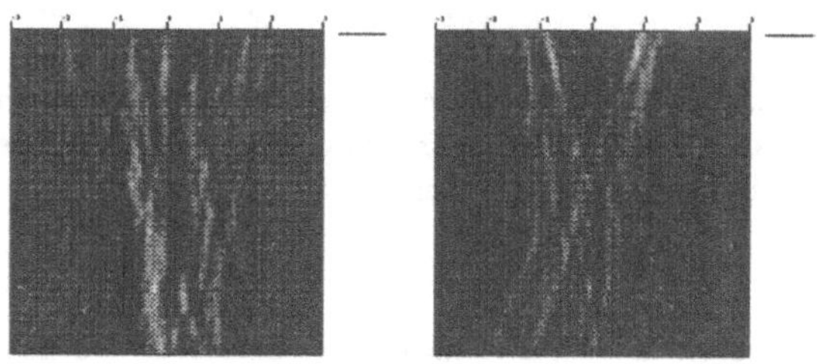

FIGURE 4.32. Result of optimizing the χ^2-Natural-Hermite index with order 5 (left picture) and χ^2-Hermite index with order 10 (right picture) for a bivariate projection.

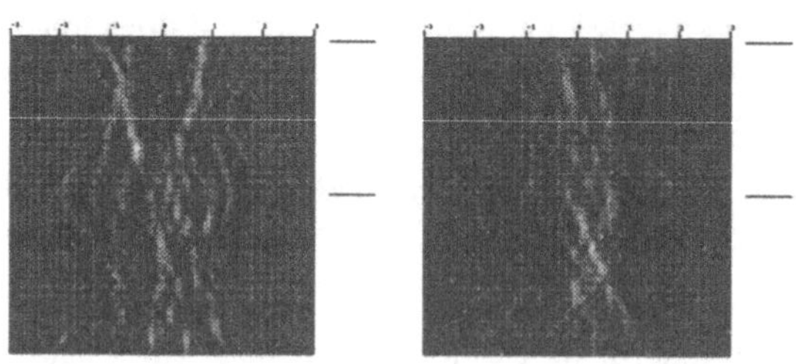

FIGURE 4.33. Result of optimizing the χ^2-Natural-Hermite index with order 5 (left picture) and χ^2-Legendre index with order 15 (right picture) for a trivariate projection.

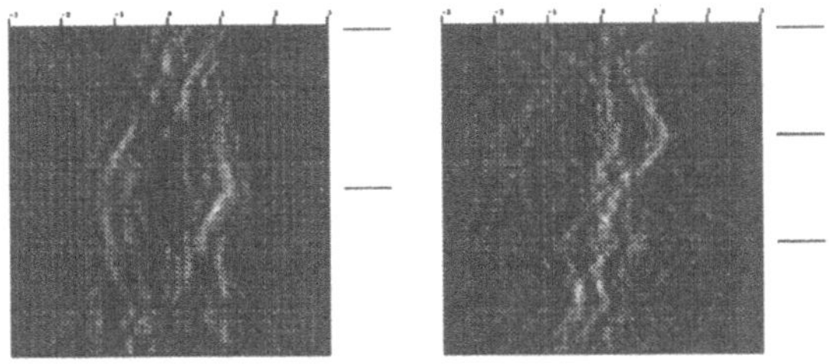

FIGURE 4.34. Result of optimizing the χ^2-Natural-Hermite index with order 15 for a trivariate projection (left picture) and χ^2-Hermite index with order 5 for a 4D-projection (right picture).

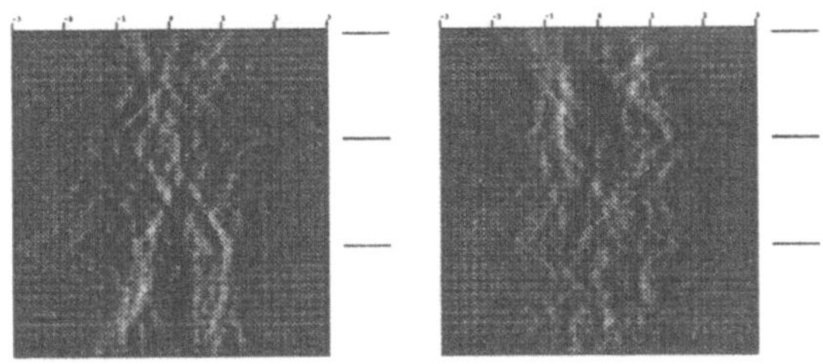

FIGURE 4.35. Result of optimizing the χ^2-Legendre index with order 15 (left picture) and order 10 (right picture) for a 4D-projection.

4.4.5 Overcoming the Problem

The fact of using the distance to $\mathbf{0}$ as a measure for the amount of structure in the projection ensures us that our indices will be rotation invariant. But if the angles between the datapoints vary the index will not notify this. So our indices will have difficulties to recognize a structure which is distributed symmetrically around $\mathbf{0}$, e.g. to distinguish three clusters from four clusters, where the radius of observation i is equal in both datasets.

One possible solution, following the idea of Yenjukov (1989), is to include an angle. We can compute the angle θ to a fixed location, e.g. e_1. Since the angle and the radius are independent we can compute an index of the form

$$\int_{[0,\infty]\times[0,2\pi]} (f_{emp}(r,\theta) - f_{gaussian}(r,\theta))^2 dr d\theta.$$

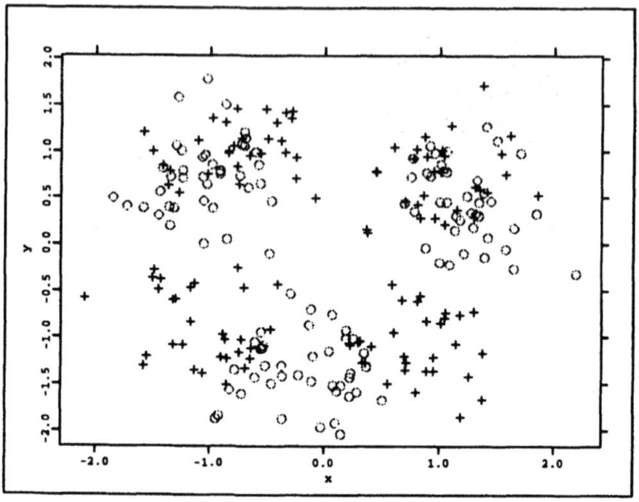

FIGURE 4.36. Data distributed in 3 cluster (marked with 'o') and 4 clusters (marked with '+').

4.5 Discrete Exploratory Projection Pursuit

In the practical work with datasets, especially in economics, it appears that we often have a mixture of continuous and noncontinuous (binary) variables. The theory developed so far has only handled continuous variables.

In the continuous case we have tried to find a projection with the largest deviation from the gaussian distribution. We have regarded the gaussian distribution as the most uninteresting. But what will be the most interesting or uninteresting distribution in the discrete case?

One possible answer would be not to take the discrete variables into account and still search the most nonnormal projection. But, as can be seen in Figures 4.38 and 4.40, a two-point distribution is a very strong nonnormal projection, which will be picked up by any of the indices. Of course the kernel based indices will pick it up very easily (see Figure 4.38).

Of course a two point distribution can not be approximated well by polynomials of low order ($J = 0, 1, 2, 3$). To approximate a two point distribution well we need at least a polynomial of order 4. Thus one might be tempted to use only low order polynomials, but this will not lead to pick up finer details of the underlying data structures as in the case of the tetrahedron dataset. Maybe we have to use mixed approximations.

Another answer could be that the uniform distribution can be regarded as

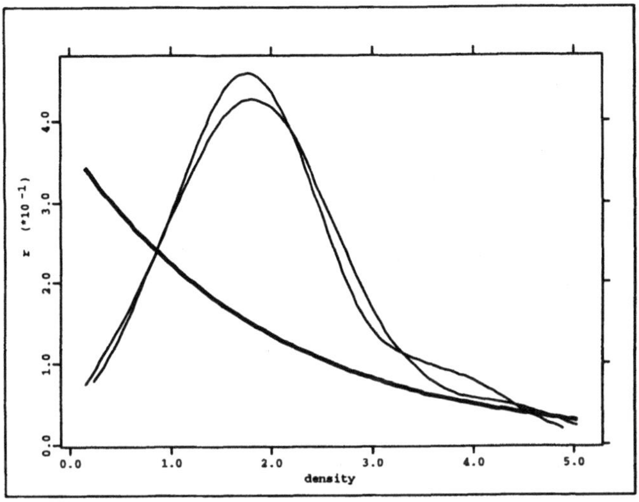

FIGURE 4.37. Kernel density estimates with rule of thumb of the datasets in Figure 4.36 (thin lines) and density of standard normal distribution (thick line).

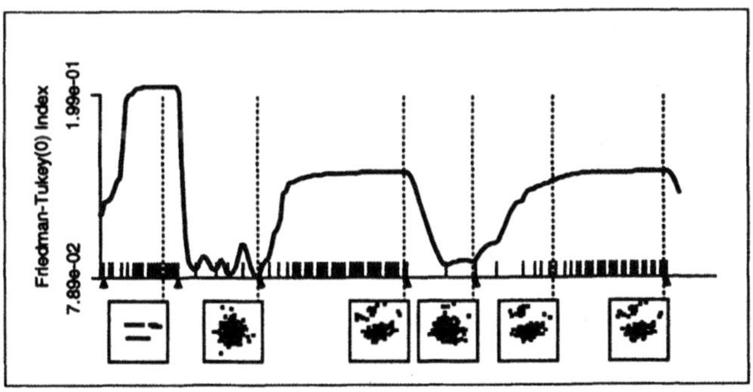

FIGURE 4.38. Friedman-Tukey index (h is the AMISE-optimal bandwidth) for the Swiss banknote dataset with a seventh variable which contains a 1 if the banknote is genuine and a 0 if the banknote is forged. We can see clearly that the two point distribution is the global maximum, but that another projection, which represents a local maximum, occurs more frequently.

the most uninteresting. In fact this will be a subjective choice which is not justified as in the case of the gaussian distribution.

In the case of exploratory projection pursuit the uninteresting normal distri-

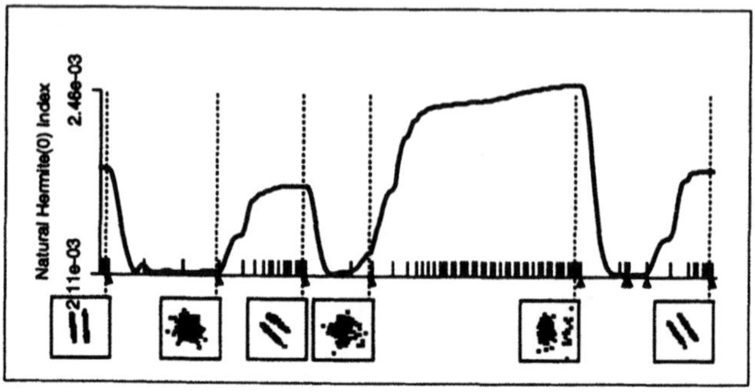

FIGURE 4.39. Natural-Hermite index ($J = 1$) for the Swiss banknote dataset with a seventh variable which contains a 1 if the banknote is genuine and a 0 if the banknote is forged. We can see clearly that the two point distribution is not the global maximum but another projection. This is due to the fact that we can not approximate the two peaks by a low order polynomial.

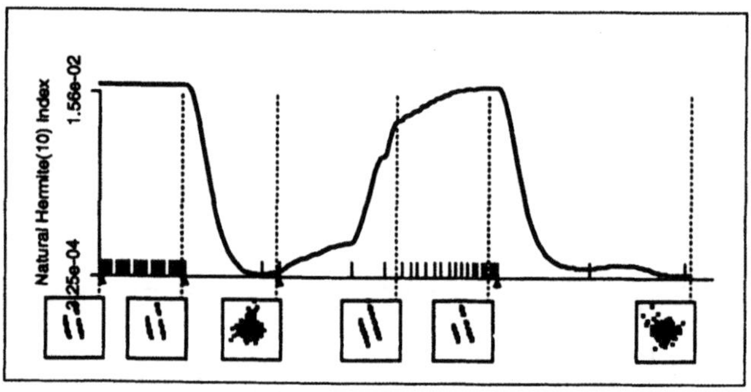

FIGURE 4.40. Natural-Hermite index ($J = 10$) for the Swiss banknote dataset with a seventh variable which contains a 1 if the banknote is genuine and a 0 if the banknote is forged. In contrast to Figure 4.39 the two point distribution now is the global maximum.

bution also represents the independence of variables since the multivariate normal distribution is the product of univariate normal distributions. In some sense a departure from the normal distribution can also mean a departure from independence. For including the discrete variables we have to find the projection which has the furthest departure from an independent distribution. Thus we would suggest the following algorithm

- Compute for each projection the density function f from the data

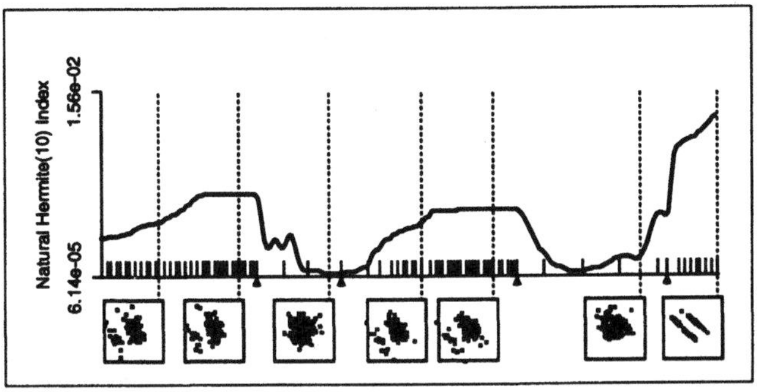

FIGURE 4.41. Natural-Hermite index ($J = 10$) for the Swiss banknote dataset with a seventh variable which contains a 1 if the banknote is genuine and a 0 if the banknote is forged. As in the case of the Friedman-Tukey index we have another projection, which represents a local maximum, and occurs more frequently.

- Compute for each projection a density function g :

$$g(y) = \int_{I\!R^{p-d}} G(y) dy_{d+1} ... dy_p$$

with $G(y) = H(Rx)$ the joint density of X_i and R a rotation matrix so that $\Gamma = (Re_1, ..., Re_d)$ are the projection vectors.

- Use an index function of the form

$$\int_{I\!R^d} (f(y) - g(y))^2 w(y) dy$$

with d the dimension of the projection, f a nonparametric estimate based on $\Gamma^T X_i$, g the density function under the assumption of independence and w as weight function (for simplicity we assume $w(y) = 1$) if necessary

- Maximize over all projections

With this method we will get the projection with the largest deviation from independence. An expansion of the unknown densities f and g with an orthonormal function xsystem $P_i(y)$ leads for g to

$$\begin{aligned} \int_{I\!R^d} g(y) P_{i_1}(y) ... P_{i_d}(y) dy_1 ... dy_d &= \int_{I\!R^p} G(y) P_{i_1}(y) ... P_{i_d}(y) dy \\ &= \int_{I\!R^p} H(x) P_{i_1}(Rx) ... P_{i_d}(Rx) dx. \end{aligned}$$

We can now include independence by decomposing $H(x) = h_1(x_1)...h_p(x_p)$:

$$\begin{aligned}
\int_{\mathbb{R}^d} g(y)P_{i_1}(y)...P_{i_d}(y)dy_1...dy_d &= \int_{\mathbb{R}^p} h_1(x_1)...h_p(x_p) \\
&\qquad P_{i_1}(Rx)...P_{i_d}(Rx)dx \\
&= \int_{\mathbb{R}^p} h_1(x_1)...h_p(x_p) \\
&\qquad \sum_{i_1+...+i_p=0}^{\infty} c_{i_1...i_p} x_1^{i_1}...x_p^{i_p} dx \\
&= \sum_{i_1+...+i_p=0}^{\infty} c_{i_1...i_p} \\
&\qquad \int_{\mathbb{R}} h_1(x_1)x_1^{i_1}dx_1 ... \int_{\mathbb{R}} h_p(x_p)x_p^{i_p}dx_p \\
&= \sum_{i_1+...+i_p=0}^{\infty} c_{i_1...i_p} E(x_1^{i_1})...E(x_p^{i_p})
\end{aligned}$$

The estimation of the coefficients for g turns out to be computationally intensive ($d = 2, J = 4$ needs more than five minutes on a PC 486DX4-100). The estimation process can not be used for interactive programs.

4.6 Requirements for a Tool Doing Exploratory Projection Pursuit

Since exploratory projection pursuit is an exploratory tool it is obvious that interactivity is needed. A tool should offer the following possibilities

- a set of index functions with the possibility to include user developed index functions
- a plot of the actual bi/multivariate projection
- a plot of the behaviour of the chosen index function in case of optimization or in case of random movement (grand tour), see Cook et al. (1995)
- recall of earlier projections

- to plot index functions on three or four dimensional subspaces, and by clicking into the function we should also get the corresponding projection
- to allow an interpretation of the found projection by linking with the variables.

5

Data Structures

Summary

In the first section we will show that graphical objects can be generated in three steps. Then we will develop a hierarchy for the graphical data structures (datapart, windows, displays). In the next section we will give reasons why matrices are no sufficient structure to store statistical data, so we need multidimensional arrays. Then we will discuss their impact on mathematical and statistical operations. The second section will close with a description why we need hierarchical objects. In the third section several forms of linking will be discussed. First we will give examples of linking plots in the thesis, then we will show further examples of linking, i.e. asking data themselves or linked data, the link of events with subroutines and at last the linking between different datasets. The fourth section will describe in short some statistical packages and indicate which features concerning data structures are available in these programs.

5.1 For Graphical Objects

5.1.1 Generating Graphical Objects

We have seen in section 2 that we have to distinguish between two kinds of graphical objects:

- points, curves and surfaces
- glyphs

In the first class we have boxplots, quantile-quantile plots, histograms, regressograms, scatterplots, 3D-scatterplots, sunflower plots, scatterplot matrices, parallel coordinate plots, Andrews curves and dendrograms. The second class consists of piecharts, star diagrams, Chernoff faces and so on.

The graphical tools of the first class can be decomposed into three steps

1. do the mathematical computation,
2. create graphical objects
3. show the graphical objects in a plot to the user.

The advantage of this method is not only that we can offer the standard tools, but we are also open for new graphical methods. The above mentioned graphics can be decomposed as follows:

Boxplot

1. Compute the 3 quartiles, the mean and the ranges.
2. Create the graphical objects like boxes, lines and the datapoints which are outside values.
3. Show the graphical objects in a plot. If we want to show several boxplots we can shift the graphical objects by an addition of a point.

Quantile-Quantile Plot

1. Compute the theoretical quantiles for one or two distributions.
2. Do nothing.
3. Show the true values and the theoretical quantiles in a plot.

Histogram, regressograms

1. Bin the data
2. Create the graphical objects (lines or outlined boxes)
3. Show the graphical objects in a plot

Scatterplot, 3D-scatterplot

1. Do nothing
2. Do nothing
3. Show the dataset in the plot and, if necessary, draw lines

Andrews curve

1. Compute a curve for every observation
2. Do nothing
3. Show the curves in a plot

Parallel coordinate plot

1. Compute a density for all axes and within axes lines
2. Do nothing
3. Show the dataset in the standard plot and combine the datapoints to curves

Tree

1. Compute a tree of merging clusters depending on the sum of the within cluster variance
2. Transform the tree into the graphical object
3. Show the dataset in the standard plot and combine the datapoints to a dendrogram

Apart from some special commands to produce the appropriate datasets, we need one type of standard plots, which basically is a 2D- or 3D-scatterplot.

5.1.2 Dataparts

In the scatterplot we want to show different datasets (e.g. in the case of the regression). This leads to the concept of dataparts which contain exactly one dataset with all the necessary attributes to plot it. The structure of the datapart will be

- data about observations
  - location of the observation
  - colour of the observation
  - size of the observation
  - form of the observation, the form can be a string
- data about lines between observations
  - which observations form one line
  - type of the line
  - colour of the line
  - thickness of the line

The structure can easily be extended to incorporate areas built up from datapoints. The shells in the cover of the book of Scott (1992) could be produced in this way.

5.1.3 Graphical Windows

A window can be composed from several dataparts. Additionally we need some parameters which are global for the window. This includes the following objects:

- the location and size of a window
- the appearance of the cursor (brush, arrow)
- the position of the cursor
- the name of the window
- the title in the window
- the projection of the data
- how many axes appear
  a 2D-scatterplot needs 2 axes, mostly with an enclosing box around, for a 3D-scatterplot or projection of higher-dimensional data we need no box, but we have to plot the axis into the data
- how the axes are scaled
  which is the minimum value, which is the maximum value, how many tickmarks, which is the output format of tickmark values
- how does an axis appear
  as a tripod in the right upper vertex of the window, as a tripod in the data with or without tickmarks, text, etc.

5.1.4 Glyph Windows

To incorporate glyphs is a difficult task. Glyph objects are piecharts, star diagrams and Chernoff faces. In du Toit, Steyn & Stumpf (1986) a lot of other glyphs like tree-diagrams are shown. We have two choices to put them into a standard 2D-window: we either use the form-parameter for a datapoint and allow a special language to produce the desired glyph, or we create an own window for this type of graphics. If we compare for example star-diagrams and Chernoff faces we see that all necessary operations like sorting, reordering etc. are the same, only the appearance changes. One disadvantage of incorporating Chernoff faces into the standard window is that the computation as given by Flury & Riedwyl (1981) is very intensive and will produce big datasets which have to be transmitted to the window. It seems reasonable to program a special window which has a parameter which tells us what the appearance will be like.

5.1.5 *Displays*

As mentioned in section 1.2.2 the flood of windows is a real problem during working. It is pointed out that we need different windows to display informations about the same task. Examples can be seen in a lot of pictures in this thesis.

So creating a display means the creation of a group of (maybe) different types of windows. Operations in one window will affect all other windows of a display.

While there is only one display visible in `XploRe` 3.2, window systems offer several displays simultaneously. The implementation in `XploRe` 4.0 will allow one display in one window, but several displays can overlap each other.

5.1.6 *Updating of Windows*

Since interactivity is necessary in statistical tasks we have to have the opportunity to influence every component of a window. The same has to be true as to the content and appearance of a window from outside. In a programming language we need a command that will allow such updates. As an example we can again use the teachware for the regression smoothing. Most of the methods incorporate a smoothing parameter which should be chosen by the user himself to see the effect of over- and undersmoothing. A change of the smoothing parameter will effect the regression line. We have to recompute the line and to plot it into the window.

5.2 For Data Objects

5.2.1 *Representation of Data as Multidimensional Arrays*

Statistical data come in two different forms to a statistician:

- as variables and
- as observations.

Statisticians have to handle both forms appropriately. In form of variables we compute a lot of statistical measurements like mean, quantiles, median, boxplots, scatterplots etc. This can lead to a storage of the data in forms of vectors such that every vector represents one variable. But for other statistical tasks we need access to the data as observations. The Chernoff faces,

Andrews curves and other techniques need an observation as their basis element. We must have access to the variables and the observations in the same manner. The solution of the problem is to represent the data in data matrices. Almost all statistical programs do it like this. The general assumption behind this representation form is that rowvectors represent observations and columnvectors represent a variable.

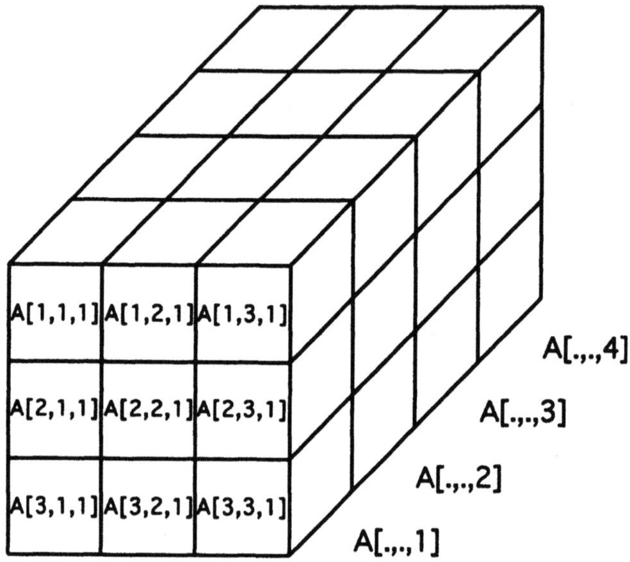

FIGURE 5.1. A three dimensional array to store for example the third moments of a multivariate dataset.

For a lot of problems this might be sufficient. But if we want to solve statistical tasks we need more than matrices.

In the section about exploratory projection pursuit we mentioned the index proposed by Jones & Sibson (1987) which is based on the third and fourth moments (see equation 4.2).

One way to implement the index could be to calculate the third and fourth moments separately for every projected dataset. Since we are looking at projections of our dataset we can establish a (simple) relationship (independent of n) between the third and fourth moment of the multivariate dataset and the third and fourth moments of the projected dataset. The third moment needs p^3, the fourth moment p^4 entries in a matrix. We can store the moments as vectors, but this would make the access to a special moment very difficult. A better solution is to create three and four dimensional arrays which contain the entries, so we need an array (for the third moment) which consists of rows, columns and layers (see Figure 5.1). This can be extended easily to four or more dimensional arrays.

Another example is local polynomial regression. As described by Fan et al. (1993) and in section 3.3.2 we have to compute for every datapoint x_0:

$$\begin{aligned}\hat{b}(x_0) &= (X^TWX)^{-1}X^TWY \\ &= S_n^{-1}(x_0)T_n(x_0)\end{aligned}$$

with

$$\begin{aligned}X &= \begin{pmatrix} 1 & (X_1 - x0) & \dots & (X_1 - x0)^p \\ \vdots & \vdots & & \vdots \\ 1 & (X_n - x0) & \dots & (X_n - x0)^p \end{pmatrix} \\ Y &= \begin{pmatrix} Y_1 \\ \vdots \\ Y_n \end{pmatrix} \\ W &= diag\left(K\left(\frac{X_i - x0}{h_n}\right)\right)\end{aligned}$$

The implementation as software for a set of k datapoints $x_1, ..., x_k$ involves the inversion of k matrices $S_n(x_l)$. Since we want to avoid looping as much as possible we would like to store all matrices $S_n(x_l)$ in data structures.

This can be done in form of a two dimensional matrix

$$S_n = \begin{pmatrix} S_n(x_1) \\ \vdots \\ S_n(x_k) \end{pmatrix}.$$

With a modified routine for the matrix inversion we can compute the inverse of the submatrices. Then we have to decompose the matrix S_n to make the necessary multiplications to compute the coefficients. Again it would be much easier to store the matrices $S_n(x_l)$ in a three-dimensional array S_n and then use the standard matrix inversion operation without having to give optional parameters.

As a consequence we need multi-dimensional arrays for to represent our data and for efficient computations. Of course we have to be aware that the use of arrays will have some administrative overhead such that array operations are slightly slower than true matrix operations.

5.2.2 Extending Mathematical Operations

It is obvious that the mathematical operations have to be extended for multidimensional operations. The following extensions are suggested:

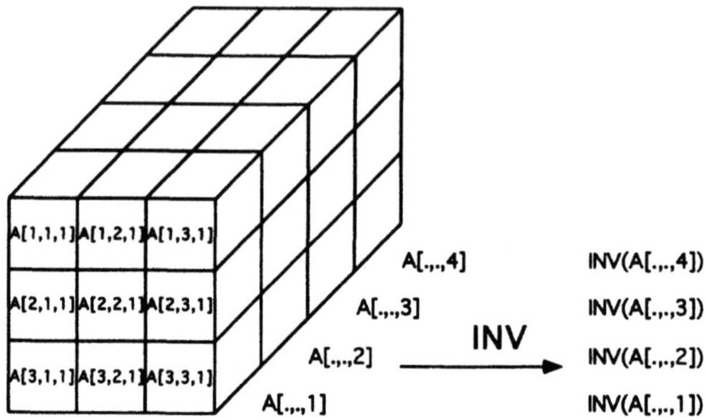

FIGURE 5.2. How the inverse matrix operation will work on a three dimensional array.

Unary mathematical functions.

Unary mathematical functions are mainly scientific functions like cosine, sine, exponential and logarithm, but also cumulative distribution functions and their inverses etc. Since these functions have only one argument and one result the extension is that we produce a multi-dimensional array of the same size as the input array, and the function is applied to each element.

Unary mathematical operators.

Unary mathematical operations are the unary minus, the logical not, the faculty and so on. These operators will also effect each element of the array and can be treated in the same way as the unary mathematical functions.

Vector operations.

We also have vector operations which are mainly operations on a variable. Typical operations are the mean, the median, the variance, the summing etc. Often it is of interest just to look at the conditional means, the conditional medians, the conditional variances, the conditional sums etc. The result will be a multi-dimensional array that in the working dimension only has one element (or k if we have k classes we condition on).

Matrix operations.

Operations which are specific for matrices are the multiplication of matrices, the inversion of matrices, the transposition of matrices, the calculation of moment matrices etc. Since the general assumption is that the rows represent the observations and the columns represent the variables, the matrix operation will work on those columns and the rows which contain one data matrix. The parallel matrices will be seen as layers so that the operation will be executed on each layer (see Figure 5.2). Optional parameters will allow to execute the operation on other layers apart from rows and columns.

Binary mathematical operators.

Binary mathematical operators are the elementwise plus, the elementwise minus and the elementwise logical comparisons. We have to take into account that we already have some kind of extension in the standard statistical languages. In EPP we need to center the data matrix by

$$Y = X - mean(X)$$

One the left side of the minus we have a $n \times p$ matrix, on the right side a $1 \times p$ vector. Thus for an elementwise operation these matrices are not compatible. Nevertheless this operation is allowed since the meaning is that we want to subtract the mean of X for each observation. As Table 5.1 shows, the programming languages do allow the elementwise operations for special sizes.

A possible generalization for a multi-dimensional array is to allow elementwise operation only if the size is the same or equal to 1 in each dimension. For two d-dimensional arrays this results in 2^{2d} valid possibilities to apply an elementwise operation.

In section 2.5.1 it is described how a histogram can be computed. In the fourth step we need to add a matrix of the form

$$\begin{pmatrix} -\delta/2 & 0 \\ +\delta/2 & 0 \end{pmatrix}$$

to each block of 2×2 matrices. This possibility can be included if we redefine the result size of an elementwise operation on two arrays. We define the size in the i-th dimension as the maximum of the sizes in the i-the dimension of the operands. With this definition we still have the possibilities given by Table 5.1, and we would be able to add blocks to the matrices as needed for the plotting of histograms. This way of dimensioning the result matrix will confuse the unexperienced user, so the first method is better as a stan-

left argument	right argument	result
$n \times p$	$n \times p$	$n \times p$
$n \times p$	$n \times 1$	$n \times p$
$n \times p$	$1 \times p$	$n \times p$
$n \times p$	1×1	$n \times p$
$n \times 1$	$n \times p$	$n \times p$
$n \times 1$	$n \times 1$	$n \times 1$
$n \times 1$	$1 \times p$	$n \times p$
$n \times 1$	1×1	$n \times 1$
$1 \times p$	$n \times p$	$n \times p$
$1 \times p$	$n \times 1$	$n \times p$
$1 \times p$	$1 \times p$	$1 \times p$
$1 \times p$	1×1	$1 \times p$
1×1	$n \times p$	$n \times 1$
1×1	$n \times 1$	$n \times 1$
1×1	$1 \times p$	$1 \times p$
1×1	1×1	1×1

TABLE 5.1. Sizes of matrices which can be used in elementwise binary operators.

dard operation. For the experienced user the second method simplifies the programing task.

Binary or n-nary mathematical functions.

Binary or n-array functions are functions like cumulative χ^2-distribution with d degrees of freedom, the normal random generator, univariate regression smoothers etc. Often we already have relationships between the parameters of a function given from the definition. A general rule to extend the parameters can not be given, but great care should be taken to find an appropriate solution of the problem. Let us take two examples:

- the normal random generator
  The form might be
  $$y = normgen(n, \mu, \Sigma)$$
  where μ is a $1 \times p \times q_1 ... \times q_k$ array and Σ is a $p \times p \times r_1 ... \times r_k$ array. In this case we can regard everything above the second dimension of the array as layers. The resulting array y would be a $n \times p \times \max(q_1, r_1) ... \times \max(q_k, r_k)$, which means we have generated in one step $\max(q_1, r_1) \times ... \times \max(q_k, r_k)$ normally distributed random samples. Since it will be a condition that $q_l = r_l$ or one of both have to be 1, we are able to

compute a lot of different random samples in one step. Especially for simulations this will be very helpful.

- the Nadaraya-Watson estimator
  The standard form of the call to compute a Nadaraya-Watson estimator will be

$$(xr, yr) = regest(x, y, bandwidth)$$

with

x	$n \times 1 \times ...$ array
y	$n \times m \times ...$ array
$bandwidth$	$1 \times 1 \times ...$ array

The result matrices will be a

xr	$k \times 1 \times max(...)$ array
yr	$k \times m \times max(...)$ array

With this definition we can compute a lot of tasks:

- a univariate Nadaraya-Watson estimator,
- a set of univariate Nadaraya-Watson estimators for different sets of y, e.g. to calculate confidence intervals and
- a set of univariate Nadaraya-Watson estimators for different sets of bandwidths, e.g. to calculate the crossvalidation function.

It is easy to build a multivariate form

$$(xr, yr) = regestp(x, y, bandwidth)$$

x	$n \times p \times ...$ array
y	$n \times m \times ...$ array
$bandwidth$	$1 \times 1 \times ...$ array
xr	$k \times p \times max(...)$ array
yr	$k \times p \times max(...)$ array

Further modifications can be done via including different kernels or a matrix of binwidths, e.g.

$$(xr, yr) = regestpkd(x, y, bandwidth, kernel, binwidth)$$

The aim of the two examples is to compute as much as possible in one step. We should keep in mind that this is just an offer to the user, we still could use loops for doing it.

5.2.3 Error Handling

It will happen very often that an error occurs in the calculation. We have to take care that the algorithm is robust against error. A stop of the whole program might not be the appropriate way to announce an error to the user. A typical example is the local polynomial regression. Assume we want to make a local polynomial regression of order 3 and the bandwidth h is chosen so small that we have at a gridpoint x less than four observations in the interval $[x - h, x + h)$. Obviously the matrix $S_n(x)$ will become singular and an inversion of the matrix will be impossible. The program needs mechanisms that first try to show the error to the user and, if the user does not care about the error, will stop the execution of the program. In the case of the local polynomial regression we might set the value for $\hat{m}(x_0)$ to a missing value.

5.2.4 Hierarchical Objects

We have shown why we need multidimensional arrays in statistics and which the implications are if we use them. Behind all the thoughts the idea is that a multi-dimensional array consists of one single type of data: either float numbers or strings or integer numbers. This is not the typical statistical object. We would like to store different kinds of objects, e.g. in economics we find a lot of datasets that consist of continuous and noncontinuous variables. We want to have names for the observations and the variables.

Another possibility is that we want to store the results of some computations with our data. The results can get lost if we store them far away from our data. The best solution is to put the result as a subobject to our data object.

We need to build up a hierarchy of objects.

Objects can consist as well of data as of programs. This leads to the concept of *object orientation*. It would be very helpful if we built statistical tools as in PPR etc.

Another advantage of hierarchical objects is that they allow to give back complex structures to the user as result of a macro. Here the computation of eigenvectors and eigenvalues, e.g. in PCA or Multidimensional Scaling can serve as examples. Another example is the result of PPR in `S-Plus` as shown in section 1.3.2.

5.3 For Linking

5.3.1 Linking Plots

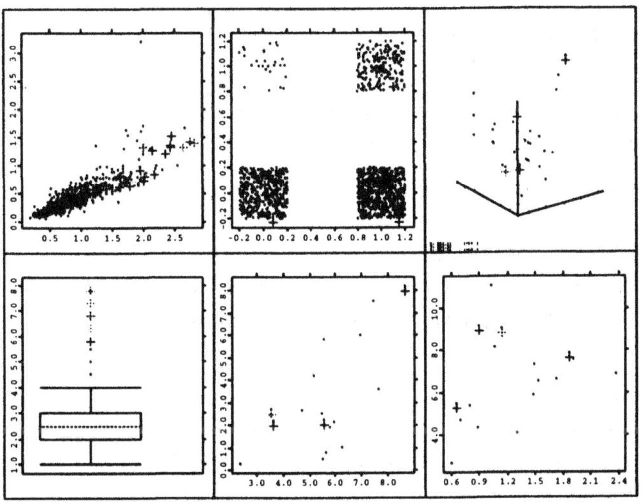

FIGURE 5.3. Different linked plots of the Berlin housing dataset (only offers from October 1994). The left upper scatterplot shows FA against FP, the central upper plot shows the variables FB against FM. Since the variables FB and FM are binary they are jittered. The lower central plot shows DU against DW so we get an impression about relationship between unemployment rate and blue-collar workers. The lower right plot shows DR against DS and will tell us something about the ecological environment. The 3D-scatterplot shows the variables DF, DU and DN. In projection pursuit regression we can interpret the first projection as an indicator of social-ecological environment, so it makes sense to have a look at these variables. The boxplot shows the variable FR; the offers with more than 5.5 rooms (6, 6.5, 7, 7.5, 8) are marked in different colours (here: gray scales).

As mentioned on page 5 the first interactive program `PRIM-9` offers some kind of linking between plots. In `PRIM-9` it is possible to make two views of one dataset (e.g. plotting X_1 against X_2 and X_3 against X_4) and to mask out some datapoints. In all plots the corresponding observations are masked out too.

This technique has been extended nowadays for all kinds of plots. We define a point-colour and a point-style for each observation. In all plots we use the observations will appear in this point-colour and this point-style. This allows us to easily identify the selected datapoints in all plots.

With the possibility of brushing (transient or nontransient) we are able to explore multivariate structures. We have already seen a lot of pictures where the linking is used:

- The linked boxplots of the variables FA and FP (Figure 2.2 on page 30)
- The subgroup analysis in the boxplots of the variable FP by FE (Figure 2.3 on page 31)
- The subgroup and regression analysis in the scatterplot of the variable FA and FP by T (Figure 2.16 on page 42)
- The brushed datapoints in a scatterplot matrix (Figure 2.22 on page 47)

Although the scatterplot matrices are well known exploratory tools, the designers of the big statistical packages now begin to integrate such tools into their programs. This is partially due to the processor speed as an interactive manipulation requires short answer times.

In all examples mentioned we always had the same type of plots available in a display. Obviously we are able to show different types of plots in a display and link them in a way that we get different views of a dataset simultaneously as in Figure 5.3.

We can learn from the Figure 5.3 that all the flats with a large number of rooms have, as expected, a large size and a high price. Because of the large size we do not expect that they are maisonette flats, but most of them have a balcony. It seems moreover that all the offers can not be found in more then 4 districts. This hypothesis can be checked by Table 5.2.

5.3.2 Linking Data

An important feature of a statistical plot is that we are able to ask data interactively about themselves. Sometimes we would like to know more about the data we see in plot. In Figure 5.4 we can move through the window and see to which dataset the observation belongs, the number of observations in the dataset and the price.

If we are able to link arbitrary informations with the data we can get many more informations about a datapoint. As an example see Figure 5.5. We have

No.	FA	FL	FR	FB	FM	FP	DF	DU	DR	DW	DN
1102	243	16	7	1	0	1450	17.5	5.5	18.6	2.3	57
1067	243	16	7	0	0	1400	17.5	5.5	18.6	2.3	57
1072	217	16	6	1	0	920	17.5	5.5	18.6	2.3	57
1071	273	16	7	1	0	1500	17.5	5.5	18.6	2.3	57
748	162	16	6	0	0	700	17.5	5.5	18.6	2.3	57
1309	235	16	6	1	1	1300	17.5	5.5	18.6	2.3	57
1232	215	16	6	1	0	1350	17.5	5.5	18.6	2.3	57
1168	172	16	6	1	0	798	17.5	5.5	18.6	2.3	57
1122	170	16	6	1	0	795	17.5	5.5	18.6	2.3	57
1125	163	16	6	0	0	895	17.5	5.5	18.6	2.3	57
1152	200	16	7	0	0	1400	17.5	5.5	18.6	2.3	57
1259	195	23	6	0	0	1000	11.8	3.6	6.5	2.2	44
1179	178	23	6	0	0	720	11.8	3.6	6.5	2.2	44
1250	186	23	6.5	1	0	798	11.8	3.6	6.5	2.2	44
1228	245	23	6	0	0	1600	11.8	3.6	6.5	2.2	44
1098	195	23	6	0	0	1000	11.8	3.6	6.5	2.2	44
1206	263	23	7.5	1	0	1400	11.8	3.6	6.5	2.2	44
1118	200	24	6	1	0	810	26.6	8.6	9.0	8.2	47
1187	280	30	8	1	0	1480	8.7	3.5	11.4	2.7	46
1147	202	30	6	0	0	890	8.7	3.5	11.4	2.7	46

TABLE 5.2. Data excerpt of the Berlin flat data. We see the offers from October 1994 with more then 5.5 rooms sorted after district numbers.

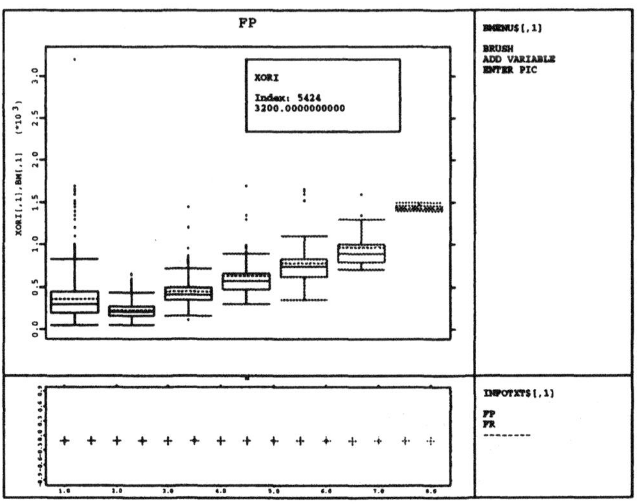

FIGURE 5.4. Boxplot of Berlin flat data (only offers from October 1994). Prices (FP) grouped by rooms (FR) plotted as boxplots. In the center of the picture we are asking a datapoint about its coordinates. We get the index and the price. The different colours are represented as gray scale.

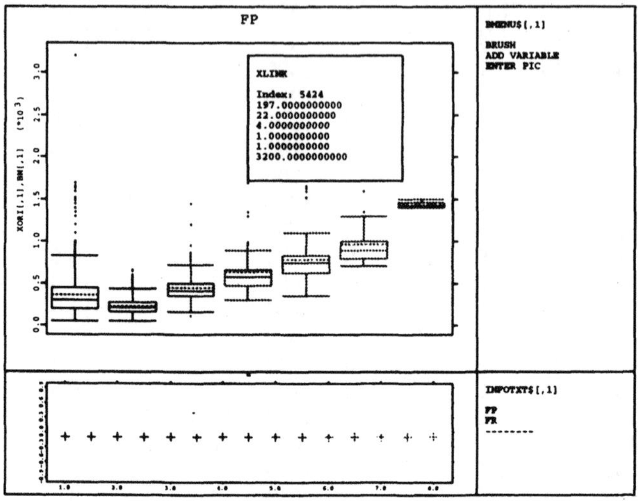

FIGURE 5.5. Boxplot of Berlin flat data (only offers from October 1994). Prices (FP) grouped by the rooms (FR) plotted as boxplots. In the center of the picture we are asking a datapoint about its coordinates. The dataset is linked to another matrix which contains more information about the flat. The different colours are represented as gray scale.

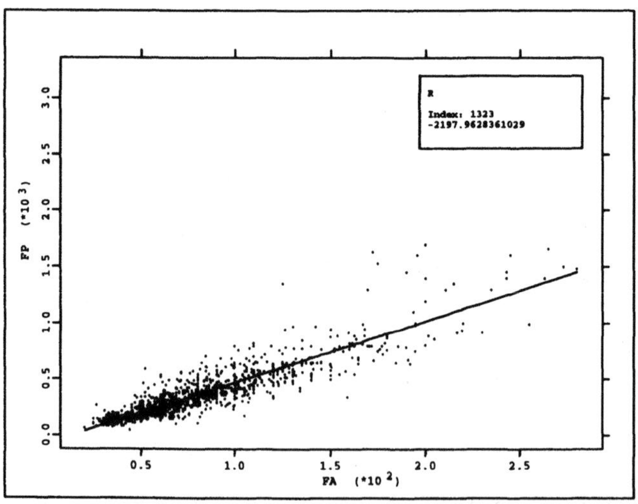

FIGURE 5.6. Linear regression of the variables FA and FP of the Berlin flat data (only offers from October 1994). The box shows the residual for one datapoint.

linked a matrix to the data with the informations about this flat. We can see the variables FA, ..., FP. We can see that the most expensive flat in October 1994 had 197 m^2, lay in Grunewald, an expensive area of Berlin, had 4 rooms and a balcony being a maisonette flat.

Another example can be found in the regression analysis. We can see the value of a residual in Figure 5.6. If we move through the dataset we can inspect all residuals.

5.3.3 Linking Events

Modern operating systems (Windows 3.1, Windows 95, OS/2, Unix) are multitasking systems. Therefore the programs have to share all the resources. Sequential programming is no longer possible. Instead of this the programs react to different events. An event can be a movement of a mouse, a mouseclick, a keypress or user-defined event.

An example can be found in Figure 2.16. Here we first brush the year of the offer and then we have to leave the picture. For each year we get one linear regression line, so from a keypress we get new informations in another window.

Another example can be found in **DataDesk**. If we compute a linear regression for a bivariate dataset we get a table of coefficients as in Figure 5.7 (here: output from SPSS). If we now move one datapoint, e.g. the observation 1323, the outlier in Figure 5.6, more to the center then the values for the linear regression will change (see the new values in Figure 5.8.

In the teachware macros of **XploRe** the same technique is used to show to the students how an outlier will influence the behaviour of the linear regression whether the L_2-error criterion or the L_1-error criterion is used.

5.3.4 Linking Datasets

The Berlin flat dataset already described consists of 14968 observations and for each observation we have all informations. But in fact the informations about the district are very often repeated. This will lead to some problems in the statistical analysis which concentrates more on the district, e.g. which is the best district to live. Even the simplest descriptive statistic, e.g. the mean of the NO_x-concentration for one time period, will produce wrong results if we use the dataset as it is.

It is clear that the collectors had the district informations available, a reconstruction for NO_x can be found in the Table E.10 revealing a strange behaviour of the data. The data are collected in a way that they are not

```
Multiple R              .88003
R Square                .77445
Adjusted R Square       .77428
Standard Error       114.39243

Analysis of Variance
                     DF        Sum of Squares        Mean Square
Regression            1        61329361.14789     61329361.14789
Residual           1365        17861883.08628        13085.62863

F =    4686.77225         Signif F =  .0000

---------------- Variables in the Equation ----------------

Variable                B        SE B       Beta          T  Sig T

FA               5.429284     .079306   .880026    68.460  .0000
(Constant)     -67.531740    6.939713               -9.731  .0000
```

FIGURE 5.7. Linear regression of the variables FA and FP of the Berlin flat data (only offers from October 1994). The observation 1323 is unchanged.

```
Multiple R              .90467
R Square                .81843
Adjusted R Square       .81830
Standard Error        97.74014

Analysis of Variance
                     DF        Sum of Squares        Mean Square
Regression            1        58779214.57938     58779214.57938
Residual           1365        13040030.20491         9553.13568

F =    6152.87132         Signif F =  .0000

---------------- Variables in the Equation ----------------

Variable                B        SE B       Beta          T  Sig T

FA               5.315207     .067761   .904673    78.440  .0000
(Constant)     -60.059451    5.929488              -10.129  .0000
```

FIGURE 5.8. Linear regression of the variables FA and FP of the Berlin flat data (only offers from October 1994). It is assumed that a misstyping occurred at the observation 1323. Now a corrected price of 1200 is used in the regression.

easy reusable for other statistical purposes. In case that we want to make an analysis about the changes in the districts over the time, we have to reduce the data.

It would be better, if the datasets were decomposed into several linked datasets, e.g. one dataset about the flats, one dataset about the districts (which can be decomposed further since some data are only collected yearly) and one time-dependend dataset. It is not clear how these single datasets could be linked together automatically.

The advantage of linked data would be more compact datasets, less storage, but it would mean a higher programming effort. The additional programming effort is due to the fact that routines like the mean woulld have to react somehow on these links.

The Berlin flat dataset consists of three groups of variables, so is reasonable to build a hierarchical object which consists of three (or more) datasets, one time-dependend dataset, the flat dataset and the district dataset. Additionally we need an object which by linking combines these datasets to one dataset. So each observation consists of three links to the three datasets.

In principle this is already done implicitly by the variables T and FL, which fix the time of the offer and the location of the flat.

5.4 Existing Computational Environments

5.4.1 Classification of Software

As already mentioned in the beginning we nowadays have a lot of statistical software available. To get an overview it will be necessary to classify statistical software. The basic classification scheme is based on the aims of the software (see Table 5.3).

But not only the aims are important, but also how many aims are satisfied.

This leads to a classification as it is used by Koch & Haag (1995) in their yearly overview of statistical software:

- Statistical software systems
  which try to satisfy a lot of aims simultaneously. Examples for these programs are `S-Plus`, `SAS`, `GAUSS` and `XploRe` 3.2.
- Special purpose programs
  which want to do just one task very well. Examples are `XGobi` and `XploRe` 2.0.

Aim	Software
online analysis	DataDesk, XGobi, XploRe 3.2
interactivity	DataDesk, XGobi, X-Lisp-Stat, XploRe 3.2
dynamic graphics	DataDesk, XGobi, X-Lisp-Stat, XploRe 3.2
linking	DataDesk, X-Lisp-Stat, XploRe 3.2
completeness	SAS, S-Plus
extensibility	GAUSS, SAS, S-Plus, X-Lisp-Stat
special topic	Mathematica, Maple V, XGobi, XploRe 2.0
teachware capabilities	XGobi, XploRe 3.2
speed	GAUSS, X-Lisp-Stat
low price	SPSS (student version)
standardization	GAUSS, S-Plus, XploRe 3.2
modern software engineering	S-Plus, XploRe 4.0
easy interface	SPSS, SYSTAT

TABLE 5.3. Examples for general aims which statistical software tries to satisfy.

- Subroutine libraries
  which can be used in other programs.
- Teachware
  which are special programs to teach statistics.

Another classification is given by user groups: Students, consultants and researches. Each has its own needs:

- Students
  Important: Low price, easy interface
  Unimportant: Extensibility, speed
  Programs: SPSS, SYSTAT, XGobi, XploRe 2.0
- Consultants
  Important: Online analysis, special topic, completeness, speed
  Unimportant: Extensibility, modern software engineering teachware capabilities,
  Programs: DataDesk, SAS,
- Researchers

Important:	Extensibility, modern software engineering, standardization, linking, dynamic graphic
Unimportant:	Easy interface, speed, low price, special topic
Programs:	**GAUSS**, **S-Plus**, **XploRe** 3.2, **X-Lisp-Stat**

As more aims a program wants to fulfill, as more work is necessary to do all the tasks. One of the big advantages of **S-Plus** is that a lot of people write additional software (see for e.g. **XGobi**) for it.

5.4.2 DataDesk

DataDesk is a software developed for MacIntosh computers. The main aim is to visualize relationships between variables. Here relationship is meant in the sense of an exploratory data analysis. **DataDesk** is only available for MacIntosh computers.

The data in **DataDesk** are stored as variables (vectors). A dataset consists of a set of variables. But **DataDesk** is only able to handle one dataset at a time. If we want to handle two datasets we have to merge the two datasets into one.

The data are linked by their index number. **DataDesk** allows linking between all graphical objects and supports subgroup analysis (brushing). It offers all graphical tools of statistics, like boxplots, scatterplots, 3D-scatterplots. But the linking is extended over the graphics. In the linear regression analysis of **DataDesk** we can make a multiple linear regression and we get the result of the linear regression as in the Figures 5.7 and 5.8 in an output window. If we now drop one of the variables in the linear regression, we will immediately get the recomputed values of the linear regression in our output window.

DataDesk offered up to version 4 no programming language, so as a consequence we can neither extend it to new algorithms nor can we see the data structure The new version 5 has now also a programming language (Theus 1996).

The help system is integrated into the help system of the MacIntosh. As the MacIntosh computers still offer a good user interface, **DataDesk** can be easily used.

Missing values are handled in the "standard" way, an observation is deleted if a missing in one variable appears. This leads to the effect that we can have a different number of points in the scatterplots when we show scatterplots of three variables.

5.4.3 GAUSS

GAUSS is a matrix-oriented programming language. It has mutated from a DOS program to a UNIX- and a DOS-program. Since **GAUSS** was mainly written in assembler which makes it very fast, the extension to UNIX took plenty of time. A lot of routines are now written in C. Nevertheless there is no perfect compatibility between DOS and UNIX. One example are the random generators. They do not give the same results on DOS and UNIX even when the same seed is used for initialization. But the UNIX-**GAUSS** offers a possibility to make the random generator to work like in DOS.

The basic data object in **GAUSS** is a matrix. The graphic possibilities of **GAUSS** in terms of interactivity are rather poor, it only offers static graphics. Since we have no dynamic graphics we have no possibility of brushing or linking.

Although **GAUSS** is a programming language it does not support any programmable menu driven environment. Nevertheless we have programs on top of **GAUSS**, e.g. MULTI, which is menu driven.

GAUSS allows a user defined error handling and supports the user by a help system. But some help, e.g. if we want to know the parameters of a command, are not easily accessible.

Mathematica and Maple V

Mathematica and **Maple V** are doing almost the same. They are no statistical programs, their emphasis is more on general mathematics. The main advantage for statisticians is the possibility of symbolic computation which allows to handle formulas.

It seems that **Mathematica** is slightly better in handling symbolic computations whereas **Maple V** gives better numerical results. **Mathematica** does not provide interactive or statistical graphics. Only the basic statistical functions are implemented (Wolfram 1991) in six standard packages:

- Descriptive statistics
- Continuous distributions
- Discrete distributions
- Hypothesis tests
- Confidence intervals
- Linear regression

Both, **Mathematica** and **Maple V**, run under different platforms. They use lists and arrays to handle objects. It is possible to build up larger units like matrices and multi-dimensional arrays.

Neither of them supports dynamic graphics, therefore we have no linking or brushing. They do not offer statistical graphics at all.

The programming language of the programs is very large, but more directed to symbolic computations. Nevertheless the handling of the programming language can be complicated. When I tried to compute the mean squared error of a kernel estimator one of the tasks was to replace a function f by its Taylor-expansion of order J. Only with some tricky handling of **Mathematica** commands I was able to achieve this.

Since both program run under a GUI all possibilities of a good help system are offered. But at least **Mathematica** has a very short help system. **Maple V** offers a topic oriented help in a way that we can click on the topics and get to see the subtopics or the appropriated command.

5.4.4 S-Plus

S-Plus is one of the latest statistical packages which runs on different platforms (UNIX, PC).

It offers a modern object orientated programming language. The object orientation supports hierarchical objects (see PPR).

Since a lot of people use it, we have a lot of routines available. The integration in a GUI environment allows good graphics although the handling could be improved. Linking and brushing are available in a limited form (scatterplot matrix), but a programmable environment for this will come soon.

The help system is not very convincing, but we have lot of literature about **S-Plus** available.

5.4.5 SAS

SAS is a old batch-orientated programming language. Nowadays it runs on a lot of different platforms. During the time **SAS** has grown to an allround program which means you will find an appropriate **SAS** function for a lot of statistical (standard) problems.

The data objects in **SAS** are variables (vectors). They can be built up to datasets (matrices). A special DATA step is used to declare these objects.

Although **SAS** now offers dynamic graphics we have only poor possibilities for brushing and linking. The programming language allows to build modules which encapsulate the whole SAS code so that the unexperienced user will only see a module where he could use a menu.

The help system is part of the GUI help system. The paper documentation of SAS is excellent, it does not only offer a description of the commands but introduces them in the topic as well.

5.4.6 SYSTAT and SPSS

SYSTAT and SPSS are menu driven programs. SYSTAT runs on MacIntosh computers and under Windows, SPSS only under Windows. We can do the standard routines under both programs.

Both programs use spreadsheets for the data input, which indicates that a matrix structure is used to store the data. But as in DataDesk both programs can only handle one dataset.

They offer mainly static graphic and some dynamic graphics (e.g. 3D-scatterplot, smoothing in scatterplots). But linking and brushing is only possible in a very limited form, e.g. SYSTAT refuses to brush more than 50 datapoints.

Although both programs have a programming language, the language is hidden. It seems to be difficult to introduce new algorithms to both program packages. A new- or redefinition of menuitems is impossible.

The help system is again integrated in the GUI help system. Nevertheless the help system is quite short. SPSS tries to introduce the user to the topic, but often the information is too sparse. Especially some of the tests used are not described sufficiently.

Missing values are again handled in the "standard" way, but at least SPSS remarks how many observations are used in the specific analysis. Yet some of the outputs are very huge (see cluster analysis) and this information can easily be overviewed.

5.4.7 X-Lisp Stat

X-Lisp-Stat is a programming language which runs on top of X-Lisp. X-Lisp itself is running on a lot of platforms. Lisp supports mainly list-like environments, and so it is possible to build up each kind of data structure we might need.

It offers all graphical possibilities of a statistical program. This includes linking, brushing and dynamic graphics. Since linking is done through Lisp it should be possible to link different datasets and to handle them appropriately.

Most of the routines are available in Lisp and we can have a look at them. The programming language allows us to build up completely menu driven environments. The help system is integrated in the GUI help system.

The main drawback from my point of view is that the programming language is based on X-Lisp. It is very difficult to change from procedural languages like Pascal, C or Fortran to Lisp which is based on the manipulation of lists.

When I tried to install `X-Lisp-Stat` for Windows I had a lot of problems which I could solve, still I was very disappointed what is being delivered together with the system. The documentation was not too good.

5.4.8 XGobi and XploRe 2.0

`XGobi` and `XploRe` 2.0 are examples of highly specialized software. `XGobi` runs only under UNIX and the main aim is to visualize multivariate data. `XploRe` 2.0 runs under DOS and is specialized in nonparametric smoothing methods. Both programs are menu driven.

The used data objects are matrices. In `XploRe` 2.0 these matrices are called "workspaces". Whereas `XploRe` 2.0 can handle several datasets (matrices), `XGobi` can handle only one dataset. But it is possible to start several `XGobi`'s which communicate with each other.

Both programs are offering dynamic and interactive graphics with the possibility of linking and brushing. It is not possible to link different graphics. In principle `XGobi` and `XploRe` 2.0 offer the possibility to extend the system which nevertheless turned out to be quite difficult. For example Prof. Schimek and his group tried to use the programming interface to extend `XploRe` 2.0, but he admitted that they had a lot of problems. I tried to include some faster EPP-indices in `XGobi` by myself, but I found it quite difficult. Finally one of the authors of `XGobi`, Dianne Cook, did it for me so that I only had to deliver the routines.

5.4.9 XploRe 3.2

`XploRe` 3.2 is an extension of `XploRe` 2.0 and runs only under DOS. It has a programming language which allows to create and use menus, though we still have a concentration on smoothing methods.

The standard data object is a matrix. In some sense it is possible to build hierarchical objects. The program offers interactive and dynamic graphics including linking and brushing. In some aspects the linking is not as good as

`X-Lisp-Stat` or `DataDesk`. Since the basis is a programming language it is more difficult to handle this program than `DataDesk`. Like `X-Lisp-Stat` it offers programmable links, but the handling is much easier. The help system is topic orientated and context-sensitive.

5.4.10 Overview About all Programs

	Data Desk	GAUSS	Mathe matica	Maple V	S-Plus	SAS	SYS TAT	SPSS	X-Lisp Stat	XGobi	XploRe 2.0	XploRe 3.2	XploRe 4.0
Multi-platform	n	y	y	y	y	y	y	n	y	n	n	n	?
Data objects													
Vector	y	y	y	y	y	y	y	y	y	y	y	y	y
Matrix	n	y	y	y	y	y	y	y	y	y	y	y	y
Matrices	n	y	y	y	y	y	n	n	y	n	y	y	y
Arrays	n	n	y	y	y	n	n	n	y	n	n	n	y
Hier. objects	n	n	n	n	y	n	n	n	y	n	n	n	y
Graphics													
Stat. graphics	y	y	y	y	y	y	y	y	y	y	y	y	y
Dyn. graphics	y	n	n	n	y	y	y	y	y	y	y	y	?
Linking													
Scatterplot matrix	y	n	n	n	y	n	y	n	y	n	y	y	?
Graph. windows	y	n	n	n	n	n	n	n	y	n	n	y	?
Diff. datasets	y	n	n	n	n	n	n	n	y	n	n	n	?
Programming language													
Menu driven	y	n	n	n	n	y	y	y	n	y	y	n	n
Available	n	y	y	y	y	y	y	y	y	n	n	y	y
Visible	n	y	y	y	y	y	n	n	y	n	n	y	y
Programable menus	n	n	n	n	n	y	n	n	y	n	n	y	?
User def. error	n	y	y	y	y	y	n	n	y	n	n	y	?
Help system													
on paper	y	y	y	y	y	y	y	y	n	y	y	y	?
online	y	y	n	y	y	y	y	y	y	y	y	y	?
topic oriented	y	y	n	y	y	y	y	y	y	n	n	y	?
context-sensitive	y	n	n	n	n	n	y	y	n	n	y	y	?
Missing treatment	y	y	y	y	y	y	n	n	y	y	y	y	?

TABLE 5.4. Comparison of the capabilities of existing statistical programs.

6

Implementation in XploRe

Summary

Here we will show the implementation of the data structures developed in the chapter before in **XploRe**. Nevertheless not everything which is explained before will be part of **XploRe** 3.2. First we will describe how graphical data structures are implemented in **XploRe** 3.2 and how they can be used interactively. Then we will show the data structures for the data being implemented in **XploRe** 3.2. The basic data structure is a matrix and no hierarchical lists are possible. Then we describe which possibilities of linking are offered in **XploRe** 3.2. Then we will describe some selected commands in **Xplore** 3.2 for producing interactivity, for reading and writing data, reading and storing macros, libraries and for binned kernel estimators. These commands will show solutions to some problems we mentioned before or which will be used to show extensions based on the extension from matrices to multivariate arrays. The third section will describe how some selected tools work (random number generator, PCA, grand tour, multidimensional scaling, clustering, multivariate kernel regression, PPR, wavelet regression, interactive contouring). It will show that we are able to implement a variety of (interactive) tools efficiently with the proposed data structures. The fourth section will describe the implementation of arrays in **XploRe** 4.0, and the fifth will show how the commands **BINDATA** and **CONV** are extended for the use with arrays.

6.1 Data Structures in XploRe 3.2

6.1.1 Graphical Objects

Displays

Creating displays. As described in section 1.2.2 a display is the largest graphical object. A display in **XploRe** is a set of nonoverlapping windows covering the whole screen. We can have several displays in **XploRe**, but only one display being active and visible. The option of various overlapping displays would have generated a lot of additional programming for a windows system.

A display can be created by the command:

```
CREATEDISPLAY (displayname, xwin ywin, type_1 ... )
```

The parameter **displayname** is the identifier for a certain display. **xwin** fixes the number of the windows in the display. If **xwin** is equal to 5 the command would create a display with $6 = 3 \times 2$ windows since 5×1 window would result in five small windows. With 6 windows we have approximately the same size in the x- and y-direction. Obviously one window is not used. The second optional parameter **ywin** allows us to fix the number of windows in the x- and y-direction, e.g. **xwin**= 5 and **ywin**= 5 would produce a display with four windows, two in each direction (see Figure 2.22). If we use negative integer values in **xwin** or **ywin**, we will get displays with asymmetric sizes of windows. For example the display in Figure 2.16 can be created with **xwin**= -2 and **ywin**= -2. The true sizes of the windows are fixed such that in x-direction the right window will be large enough to show one value of one vector without cutting decimals.

The last parameter **type_1 ...** fixes the type of the window (see below). If the sequence of types can be repeated it is only necessary to give the repeating sequence once. We do not have the freedom to give names to the windows as they will be named automatically. So if we have three two dimensional windows and one text window as in Figure 2.16, the names will be **s2d1**, **s2d2**, **s2d3** and **text1**.

Standard displays

We have several standard displays with one window in **XploRe** corresponding to the possible window type:

- A help display which is used for displaying a helptext,
- a text display which is used for displaying various texts,
- a static2d display for showing 2D-scatterplots,
- a dynamic3d display for showing 3D-scatterplots,
- a face display for showing Chernoff faces and
- a boxplot display for showing boxplots.

In contrast to the window types above the histogram of the three-step model will follow as described in section 5.1.1, using the static2d display.

Changing and killing displays. To change from one display to the other we have the command

```
DISPLAY (displayname).
```

Since displays are global objects in **XploRe** it is necessary to kill them explicitly with the **FREE** command.

Printing displays. A display can be printed when we are in the active window by the key combination `<Alt-p>` or by the command `PRINT`. In `XploRe` it is impossible to print one window only.

Internal structure. The internal structure of a display is:

```
struct display
{ struct window *firstwindow, *workwindow;
  int freewind, fullwind;
};
```

We have pointers to two windows: the first window of the display and the active window. The active window is the one where the cursor is located. The variables `freewind` and `fullwind` give us the number of windows used and not used in the display. The structure of the window will produce the queue of all windows in a display. For easier access we have a pointer to the active window.

Windows

Window types. As mentioned before we have five different types of windows:

- text windows
- 2D-scatterplot windows
- 3D-scatterplot windows
- Boxplot windows
- Chernoff face windows

Windows are filled with data by the command

```
SHOW (data_1 ... data_p window)
```

Only in the case of the Chernoff face window there is a deviation; the command then is

```
SHOW (data xface yface index type).
```

Text. The text windows are used to display and manipulate text. The editor of `XploRe` uses such windows for editing data, text and programs, the command is `EDIT`. The command `MENU` allows linewise selection with the cursor.

2D-scatterplot. For displaying two dimensional data the 2D-scatterplot window is used. A dataset `data_1` can contain more than two vectors. In this case we will see the first vector plotted against the second, the third against the fourth and so on. We can enter several datasets (`data_2` ...), and each dataset will get a different colour, see Table 6.1. The true colour on the screen is depending on the manipulation of the internal colour palette by the command `PALETTE`.

Dataset no.	Colour
1	White
2	Yellow
3	Light magenta
4	Light red
5	Light cyan
6	Light green
7	Light blue
8	Dark gray
9	Light gray
10	Brown
11	Magenta
12	Red
13	Cyan
14	Green
15	Blue
16	Black

TABLE 6.1. Colours of datapoints in the i-th dataset. The coding of the colours follows the standard scheme given by DOS. With the command `PALETTE` a complete recoding of the colours is possible, since e.g. yellow is not always a good colour for plotting.

Since we want to have interactive graphics which means we want to modify the appearance of the graphic we have an icon bar for the active window in the right upper corner of the screen. It can be accessed via the cursor and `<Enter>` or the function keys (`<F1>` - `<F10>`). The icons, each of them hiding a menu, have the following meaning:

F1 Help

F3 Go into link modus (details see Section 6.1.3)

F4 Change window attributes

- **F1** Help
- **F2** Headline on/off
- **F3** Axis on/off
- **F4** Edit headline and axis text

- **F5** Exit axis borders (minimum, maximum, tickwidth)
- **F10** Return to main icon bar

F5 Brushing operation

- **F1** Help
- **F2** Zoom in brush area (focussing)
- **F3** Zoom out brush area
- **F4** Change brush size
- **F5** Brush on/off
- **F6** Select colour for brushing
- **F7** Select point-style for brushing
- **F10** Return to main icon bar

F8 Activate the following window in the display

F9 Activate the preceding window in the display

F10 Change datastyles

- **F1** Help
- **F2** Change the colour of the active dataset
- **F3** Change the point-style of the active dataset
- **F4** Lines on/off
- **F5** Change the line style
- **F6** Change the line thickness
- **F7** Select the data for linking (details see Section 6.1.3
- **Enter** Select the active dataset
- **Esc** Return to main icon bar

Here we notice an inconsistency in the selection of colours and point-styles for brushing and changing datastyles. In general we should use the same hotkeys for the same tasks.

Two subicon bars will need a more detailed explanation: brushing operations and changing datastyles.

If we press `<F5>` the cursor will change from a small arrow to a rectangular box. The starting size is depending on the windowsize. In contrast to the brushing modes described, `XploRe` has only the transient brushing mode. We just need to toggle the brushing on and off with `<F5>`. Additionally we have the possibility of focussing (see Figure 2.13 and 2.14).

If we press <F10> a box is opened which contains all datasets from `data1` to `datap` as given in the `SHOW` command. We can change the appearance separately for each datapart. We just have to move the cursor on the line where we find `datak` and press <Enter>. Now this datapart is activated and the actual parameters (colour, point-style etc.) are shown in the icon bar, and we can manipulate the datapart by pressing the function keys.

3D-scatterplot. The 3D-scatterplot is very similar to the 2D-scatterplot, just the appearance of the axes is different. Again a dataset can have more than three vectors. So the first, second, third and the fourth, fifth and sixth etc are plotted together. But this structure has a deeper meaning. Sometimes we need to plot a surface in 3D-space, so a dataset with $3 \times m$ vectors is interpreted as a grid (x_i, y_j, z_{ij}). If we draw lines instead of datapoints we will get a grid in the space instead of wildly connected datapoints. The `SPLIT` command allows easily to split a dataset with three vectors into a dataset with $3 \times m$ vectors.

Obviously we need more features for the 3D-scatterplot. The icon bar offers additionally:

F4 Change window attributes

- **F6** Edit rotation origin

F6 Rotation

- **F1** Help
- **F2** Go into rotating mode (cursor left, right, up and down and cursor control left and control right rotate the data)
- **F3** Move closer to or further from the data
- **F4** Change view direction (standard view is to the "center" of the data)
- **F5** Scale axis
- **F6** Select projection 1
- **F7** Select projection 2
- **F8** Select projection 3
- **F9** Select projection 4
- **F10** Return to main icon bar

F7 Reset to start projection

Boxplot

The boxplot window displays each vector of a dataset as a boxplot. Here we have one optional parameter which fixes the colour and the point-style of the boxplots:

```
SHOW (data boxes type).
```

Chernoff faces

The Chernoff face window displays each observation of a multivariate dataset as a Chernoff face (the Flury-Riedwyl algorithm is used). The command differs from the standard show command:

```
SHOW (data xface yface index type).
```

The parameters **xface** and **yface** fix how many faces will be shown in the x- and y-direction. The index fixes which vector in the dataset is represented by which face part. The maximum number of face parts is limited to 36:

1	Right eye size,
2	right pupil size,
3	position of right pupil,
4	right eye slant,
5	horizontal position of right eye,
6	vertical position of right eye,
7	curvature of right eyebrow,
8	density of right eyebrow,
9	horizontal position of right eye brow,
10	vertical position of right eye brow,
11	right upper hair line,
12	right lower hair line,
13	right face line,
14	darkness of right hair,
15	right hair slant,
16	right nose line,
17	right size of mouth,
18	right curvature of mouth,
19-36	the same as 1-18 but for the left part of the face

For example if **index[2,1]**= 5 then the fifth vector in **data** is used to determine the right pupil size. If the **index** vector can only have 18 entries, it is assumed then that the faces are symmetrical.

Manipulating windows

In **XploRe** we have two possibilities to manipulate windows. The first possibility allows us to insert keystrokes via **WRITECON** into the keybuffer. These inserts appear in the window when the a user has pressed them. An example often used in **XploRe** macros is:

```
WRITECON(27)
SHOW(x S2D)
```

We insert <Esc> (= ASCII 27) into the keybuffer. Normally we will stay in the S2D-window when SHOW is executed. But the program will find <Esc> in the keybuffer and exit from the window immediately. If a user would like to leave the window he would have to press <Esc> too. A whole sequence of keypresses can be entered in the keybuffer, e.g. see the macro CONT3D.

The other possibility offers the UPDATE command. The parameters are

```
z = UPDATE (x {n {s1 {s2 {...}}}}).
```

The first parameter is a dataset which can be appended to the existing datasets or which can replace an existing dataset. The second parameter fixes the number of dataparts. If in the SHOW command two datasets were given and n is equal to two then the dataset x will replace the second datapart. If the number is larger than the number of existing dataparts, the dataset will be appended. To find out which is the number (position) of the new datapart UPDATE returns z. All other parameters s1, s2, ... are used to manipulate the attributes of a datapart or the attributes of the window. Some of the parameters are only keywords which need additional parameters:

- S2Di, D3Di, ... will activate another window of a display
- LINE, POINT draw the specified dataset as lines or points
- SOLID, DOTTED, CENTER, DASHED fix the line style
- THICK, NORM fix the thickness of the line
- TITLE "string" or string matrix which changes the title of the active window
- XAXIS "string" or string matrix which changes the x-axis text of the active window
- YAXIS "string" or string matrix which changes the y-axis text of the active window
- ZAXIS "string" or string matrix which changes the z-axis text of the active window
- OLDPROJ keeps the actual projection even if new data are inserted into the window
- NEWPROJ computes a new projection from all data

Internal structure

The internal structure is quite complicated. The window has to know a lot of data about itself:

```
struct window
{ int xmin, ymin, xmax, ymax, bwidth, bheight,
      gxmin, gymin, gxmax, gymax,
      free, type, flags, new, cursorx, cursory,
      movex, movey, limitx, limity,
      faceno;
  char *winname, *namexaxis, *nameyaxis, *namezaxis, *namehead,
       bmode, axistextmode;
  struct datapart *firstdata;
  struct window *nextwin, *lastwin;
  projection *proj;
  double extrema[12], middle[3], windstyle[MAXWINDOWSTYLE];
};
```

The parameters have the following meaning:

- **xmin**, **ymin**, **xmax** and **ymax**
  Position of the window in the display in screen coordinates

- **bwidth**, **bheight**
  Height and width of the brush

- **bmode**
  The colour and the style a datapoint gets if **<F5>** is pressed in the brushing icon bar

- **gxmin**, **gymin**, **gxmax**, **gymax**
  Position of the drawing area

- **free**
  Is the window used or not

- **type**
  Type of the window (2D-scatterplot, 3D-scatterplot, ...)

- **flags**
  Flags to indicate if a frame is to be drawn etc.

- **new**
  Distinguishes between the three window drawing modes

- **cursorx**, **cursory**
  Position of the cursor

- **movex**, **movey**
  Pixels to move if the next left, right, up or down key is pressed

- **limitx**, **limity**
  Limits of the cursor movement
- **faceno**
  First observations which will be drawn in the upper left corner
- **winname**
  Name of the window (e.g. S2D1)
- **namexaxis**, **nameyaxis**, **namezaxis**
  Axes text
- **namehead**
  Headline text
- **axistextmode**
  Axes mode; if the window gets too small it does not make sense to draw axes, tickmarks and axes texts
- **firstdata**
  Pointer to the first datapart
- **nextwin**, **lastwin**
  Pointers to the preceding and the following window in the display
- **proj**
  Projection matrix
- **extrema[12]**
  Extremal values in all three directions
- **middle[3]**
  Central coordinates, e.g. for rotating
- **windstyle[MAXWINDOWSTYLE]**
  Flags for the appearance of the window (Headline on/off, Axes on/off etc.)

Datapart

A datapart is each dataset given to a window through a **SHOW** or **UPDATE** command. The number of dataparts is not limited in **XploRe**. The internal structure is given by

```
struct datapart
{ char *firstmatrix;
  char *linkmatrix;
  int *screenx, *screeny, screendim, screenlen;
```

```
  char *screentext;
  double datastyle[MAXDATASTYLE];
  struct datapart *nextdata;
};
```

and the meaning of the parameters is

- **firstmatrix**
  The name of the data matrix
- **linkmatrix**
  The name of the linked matrix or the linked datapart
- **screenx**, **screeny**
  The screen coordinates for each datapoint
- **screendim**, **screenlen**
  The dimension and the length of the data matrix
- **screentext**
  The text which should be plotted instead of a point
- **datastyle[MAXDATASTYLE]**
  Information how the dataset should be plotted
- **nextdata**
  Pointer to the next datapart in the queue

For fast redrawing the screen coordinates are kept in vectors, e.g. if we call **DISPLAY**. For the matrix and the link object we only keep the names but no copies of the matrices themselves. This causes problems if we use temporary objects, but we can exchange a data matrix, and it only requires a simple redraw to show the new data.

6.1.2 Data Objects

Vectors. The smallest data objects are vectors in **XploRe** 3.2. This has partly historical reasons, because **XploRe** 2.0 has used vectors, but it allows us to handle vectors for different types. From vectors of different types we build up matrices. All vectors in a matrix have to have the same length.

In **XploRe** has the following vector types:

- Float
  The vector type is used to store 8-byte float numbers

- Text
  The vector type is used to store strings
- Mask
  The vector type is used to store point-colours and point-styles
- List
  The vector type is used to store lists of matrices

The list vectors are mainly used internally, e.g. a function call like

```
(xb yb) = BINDATA(x binwidth origin y)
```

will build up a list vector which contains the elements **x**, **binwidth**, **origin** and **y**, so each command in **XploRe** has (internally) only one input parameter. The same happens with the return parameter of **BINDATA**. We get back a list vector with the two elements **xb** and **yb**.

Internal structure. As consequence the internal structures looks like:

```
struct matrix
{ int dim, len;
  struct vector *firstvec;
};
```

The parameter **dim** and **len** describe the size of the matrix. **firstvec** is the address of the first vector.

```
struct vector
{ int bufsize;
  int noofmiss;
  double min,max;
  char type;
  char *name;
  struct vector *nextvec;
  void *buffer;
};
```

The vector structure is larger. **type** fixes the type of the vector, **nextvec** points to the next vector of the matrix. **name** contains the name of the vector, in general *matrixname[,i]*. We store additionally the number of missing values, the minimum and the maximum value in **min** and **max**. **buffer** is the address where the data are really stored in the memory, **bufsize** indicates how much memory in bytes is used.

The object list

```
struct object
{ char *name, *origin;
  int type, depth;
  void *ptr;
  struct object *next;
  struct object *preobj;
};
```

The object list which holds all objects (matrices, macros, displays) in `XploRe` is a doubled linked list. The parameter have the meaning:

- `name`
  Name of the object, e.g. matrixname, procedurename, displayname
- `origin`
  If an object is a macro then the filename is stored so that the macro can be easily loaded
- `type`
  Type of the object
- `depth`
  The depth of the object. It tells where the object was generated: 0 - from the commandline, 1 - from a macro called from the commandline, 2 - from a macro called from a macro which was called from the commandline and so on
- `ptr`
  Address of the content of the object
- `next`, `preobj`
  Addresses of the following and preceding object in the object list

Although we have a flat list structure for the object list we are able (in general) to build hierarchical objects by the use of list vectors.

6.1.3 Linking Objects

Interactive linking. In `XploRe` we have several possibilities to create links between a datapart in a window and other dataparts or matrices.

As mentioned on page 201 we can press `<F10>` in a graphical window. We then get a box with all dataparts. If we now press `<F7>` we get another box with all matrices which have a length conforming to the datapart. Now we can link a matrix or a datapart to this datapart.

When we are back in the graphical window we can toggle to the link modus (`<F3>`). A box appears with the coordinates of the datapoint, if we have not linked a matrix to the datapart. If we have linked a matrix to the datapart then the corresponding (through the number of the datapoint) values of the matrix will be shown. So it is possible to plot a dataset, a regression line and to ask the data about the residuals as in Figure 5.6.

Linking by command

In `XploRe` there is the `LINK` command. It allows to link dataparts with other dataparts or matrices. The syntax is:

```
LINK (name1 name2).
```

The parameter `name1` is the name of a datapart of a window in the active display, e.g. `s2d1data_1`. `name2` can be a datapartname or a matrixname (e.g. `x`). It is not possible to link between dataparts of various displays, since we can only show one display on the screen which will change in `XploRe` 4.0. If we link a matrix to a datapart this is of any influence only if we press `<F3>` in a graphical window. If we link another datapart the brushing influences it. If we brush datapoints of a datapart the corresponding datapoints in the linked datapart we will get the same point-colour and point-style. The scatterplot matrix is a good example in `XploRe` (see Figure 2.22).

6.2 Selected Commands in XploRe 3.2

6.2.1 Interactivity

Interactivity is necessary in a lot of statistical applications. We have already described the interactive graphics in `XploRe`. Nevertheless we need more than just linking pictures and changing graphics. The system has to react on user input.

The input commands are **READCON**, **READVAL**, **READSTR** and **MENU**:

```
z = READCON (x {filter})
z = READVAL (x {def})
z = READSTR (x)
pos = MENU (x {window})
```

READCON reads the next keypress from the internal keybuffer. The optional parameter `filter` allows to filter characters as all characters will be ignored which are not in the `filter`. The concept of a `XploRe`-internal keybuffer allows automatization of keypresses. With the command **WRITECON** we can insert keypresses in the internal keybuffer. This leads to sequences of **WRITECON**'s, e.g. in the interactive contouring.

READVAL asks for a number via a messagebox from the user. The parameter **x** is a message to the user and the optional parameter **def** a default value, if the user leaves the box without typing a number. **READSTR** reads a string from the user.

MENU shows a menu in a text-window of the active display and allows the user to choose. The parameter **x** contains the text of the menu. The output-parameter **pos** contains the number of the selected line and the line itself.

6.2.2 Reading and Writing Data

As mentioned earlier the reading and writing of data can be a painful task. Here the reading and writing of data is based on ASCII-files. The command

```
x = READ ("bank2")
```

will read the Swiss banknote dataset into the matrix **x**. Since in each line of **bank2.dat** we have exactly 6 numbers and 200 lines, it follows that **x** has the dimension 200×6. The writing is as easy as reading:

```
WRITE (x "bank2")
```

and we write our data to the file **bank2.dat** (and overwrite the file).

If we have multidimensional arrays as in **XploRe** 4.0 instead of matrices, we can not keep the dimensions larger than 2. An array of the dimension $n_1 \times n_2 \times n_3 \times ... \times n_p$ will be written in $n_1 \times n_3 \times ... \times n_p$ lines and n_2 columns. If we read these data with **READ** it will result in the appropriate matrix. With the help of the **RESHAPE** command in **XploRe** 4.0 we can easily reconstruct the original array, if we know the dimensions.

Sometimes we want to store the dimensions too. **XploRe** has a X-format, an internal format to store matrices. The reason for the introduction of the X-format in **XploRe** 3.2 was that large datasets need a lot of time to be loaded. In the X-format reading is much faster as the internal structures can be stored directly. The **READ** command recognizes automatically if a dataset is in ASCII-format or in X-format. Only the **WRITE** command needs a special format parameter to store datasets in X-format:

WRITE(x "xbank2" "X") with compression

or

WRITE(x "xbank2" "Y") without compression.

6.2.3 Random Generators

XploRe 3.2 provides two commands for the random generators:

- **UNIFORM** which fills an $n \times p$ matrix with uniform random numbers in $[0, 1]$ and
- **NORMAL** which fills an $n \times p$ matrix with normal random numbers $N(0, 1)$.

The normal random generator is built via the Box-Muller algorithm from the uniform random generator; see Press, Flannery, Teukolsky & Vetterling (1988).

The uniform random generator is a linear congruential random generator. The value r_i will be computed from r_{i-1} via

$$r_i = \frac{(27132 * r_{i-1} + 7.0) \equiv 62748517}{62748517}.$$

6.2.4 FUNC and LIBRARY

XploRe offers two possibilities to load functions. The command **FUNC** loads a single macro from a file:

FUNC (filename).

It is expected, but not necessary, that the macro has the same name as the file. The method is mainly used during the development of a macro for a library or in library macros to load additional macros.

The command **LIBRARY** loads a whole set of macros from a file:

LIBRARY (filename).

Each line of the file contains the names of the macros. An example is the **HIGHDIM**-library:

```
; ***************************************************************
; ********* the XploRe  HIGHDIM library  ************************
; ***************************************************************
; ************ date   : 950725 **********************************
; ************ author: Sigbert Klinke ***************************
; ***************************************************************
DRAFTMAN          ; Draftman plot
 DRAFT4           ;    for  4D-data
 DRAFT5           ;    for  5D-data
 DRAFT6           ;    for  6D-data
```

```
 DRAFT7          ;   for  7D-data
 DRAFTINF        ;   for  >7D-data
DOFACE           ; Flury faces plot
SIMDEP           ; Simplicial depth
TOUR             ; Grand tour
SUNFLOW          ; Sunflower plot
DENS3PLT         ; Density estimation for 3D-dataset
ANDREW           ; Andrews curve
JITTER           ; Jittering of data
CONTOUR          ; Contouring of 3 or 4 dimensional data
MULTDRAW         ; Shows movements of datapoints in 2/3-space
PCPLOT           ; Parallel coordinates plots
CLUSTER          ; Explorative cluster analysis
HDPCA            ; Principal component analysis
BOXCOMP          ; Linked boxplots
BOXSUB           ; Subgrouped boxplots
VARCOMP          ; Univariate analogue to draftman plot
SCATTER2         ; Analysis of scatterplot
DESKSTAT         ; Descriptive statistics
DOMDS            ; Show MDS results
MDS              ; Makes metric multidimensional scaling
NMDS             ; Makes nonmetric multidimensional scaling
```

Since the directory structure of **XploRe** is fixed, we know where to find the macros.

The **HIGHDIM** library contains techniques which are related to the analysis of multidimensional datasets.

6.2.5 *BINDATA and CONV*

The important commands for every method connected to kernels are the commands **BINDATA** and **CONV**. The following macros of **XploRe** use these commands:

- **ASH.XPL**, **ASHK.XPL**, **HISTOGRM.XPL**, **SUNFLOW.XPL**
  Average shifted histograms, histograms, sunflower plots
- **BWSEL.XPL**, **HSJMBIN.XPL**
  Bandwidth selection methods for kernel density estimation
- **DENAUTO.XPL**, **DENEST.XPL**, **DENEST2.XPL**, **DENGAU.XPL**, **TWDENES1.XPL**, **TWDENES2.XPL**
  Kernel density estimators

- **DENS3PLT.XPL**
  Interactive plotting of marginal densities of a 3D-dataset
- **CONT2.XPL, CONT3.XPL**
  Interactive contouring routine for 3D- and 4D-datasets
- **LPD1EST.XPL, LPDEREST.XPL, LPREGEST.XPL**
  Local polynomial estimation
- **REGAUTO.XPL, REGEST.XPL, REGLNBC.XPL, REGMHFH.XPL, TWKERLIK.XPL, TWKERREG.XPL, TWLSCRVL.XPL, TWRECRVL.XPL, TWREGEST.XPL**
  Kernel regression
- **TWWVLET.XPL, WAVEREG.XPL**
  Wavelet regression
- **CONTING.XPL, PPINTER.XPL, SCATTER2.XPL**
  Different algorithms in cluster analysis, exploratory projection pursuit and 2D-dataset analysis

The syntax of the **BINDATA** command is

```
(xb yb)=bindata(x d o y).
```

The parameter **x** is a multivariate dataset, **d** is the binwidth, **o** the origin of the bin 0 and **y** the (optional) y-values to **x**. The result **xb** will contain the bin position of the nonempty bins as integers. **yb** contains in the first columns the number of observations which have fallen into the bin. The second and more columns contain the sum of the y's which have fallen into the bin.

The real power concerning kernel estimation comes from the interplay with **CONV**. The call is

```
(xc yc or)=CONV(xb yb wx wy sym).
```

CONV computes for each gridpoint $(x_{i_1,1}, ..., x_{i_p,p})$ in the grid $\min(\mathtt{xb}[,1]) \leq i_1 \leq \max(\mathtt{xb}[,1]), ..., \min(\mathtt{xb}[,\mathtt{p}]) \leq i_p \leq \max(\mathtt{xb}[,\mathtt{p}])$ the function

$$\sum_{i=1}^{n} \mathtt{yb}[\mathtt{i},] * \mathtt{wy}[\mathtt{abs}(i_1 - \mathtt{xb}[\mathtt{i},1]), ..., \mathtt{abs}(i_p - \mathtt{xb}[\mathtt{i},\mathtt{p}])].$$

Assume we generate 10 uniform values

$$x = (1.11e-007, 0.0030, 0.12, 0.42, 0.0035, 0.23, 0.42, 0.60, 0.61, 0.26)$$

and bin them with $d = 0.2$ and $o = 0$ and we get

$$\begin{aligned} xb &= (0,2,1,3) \\ yb &= (4,2,2,2). \end{aligned}$$

To estimate the kernel density with the triangle kernel and a bandwidth of 0.4 we can create a binned version of the kernel (with $d = 0.2$ and $o = 0$) and we get

$$\begin{aligned} wx &= (-2,-1,0,1,2) \\ wy &= (0,1/4,1/2,1/4,0). \end{aligned}$$

Since the kernels are always symmetrical, for computational efficiency we only store the positive part and omit the zeros. We will have

$$\begin{aligned} wx &= (0,1) \\ wy &= (1/2,1/4). \end{aligned}$$

These are the parameters which **CONV** needs to work:

```
(xc yc or) = CONV (xb yb wx wy).
```

Now the kernel estimation sets the kernel at each position X_i and sums up. Since we have in **xb** four observations, **CONV** needs exactly four steps. First we generate **xc** from min(**xb**) to max(**xb**) and fill **yc** with zeros and then we move the kernel step by step over the data; see Table 6.2 as an example.

step		1		2		3		4	
xb		0		2		1		3	
yb		4		2		2		2	
xc	yc	wy	yc	wy	yc	wy	yc	wy	yc
0	0.0	1/2	2.0	0	2.0	1/4	2.5	0	2.5
1	0.0	1/4	1.0	1/4	1.5	1/2	2.5	0	2.5
2	0.0	0	0.0	1/2	1.0	1/4	1.5	1/4	2.0
3	0.0	0	0.0	1/4	0.5	0	0.5	1/2	1.5

TABLE 6.2. How **CONV** works for a function.

The same technique can be used for the multivariate case. Sometimes it is interesting to compute the derivative of the kernel. This can be done by using the derivative of the kernel in the density estimation. The symmetry of the kernel $K(-x) = K(x)$ changes to $K'(-x) = -K'(x)$. The optional parameter **sym** allows putting in such origin symmetric kernels by setting **sym** to one.

In the trivariate case we calculate the partial derivatives $\partial f / \partial x_{i=1,2,3}$ by composing **sym** as

$$\begin{array}{llll} \texttt{sym} & = 1 = & 1 * 2^0 + 0 * 2^1 + 0 * 2^2 & \text{for } \partial f / \partial x_1 \\ \texttt{sym} & = 2 = & 0 * 2^0 + 1 * 2^1 + 0 * 2^2 & \text{for } \partial f / \partial x_2 \\ \texttt{sym} & = 4 = & 0 * 2^0 + 0 * 2^1 + 1 * 2^2 & \text{for } \partial f / \partial x_3. \end{array}$$

We tell **CONV** by **sym** which kind of asymmetry we need. For the univariate derivative we have

$$\begin{aligned} wx &= (0,1) \\ wy &= (0,-1). \end{aligned}$$

The call is

```
(xc yc or)=conv(xb yb wx wy 1)
```

and the work of **CONV** is shown in Table 6.3.

step		1		2		3		4	
xb		0		2		1		3	
yb		4		2		2		2	
xc	**yc**	**wy**	**yc**	**wy**	**yc**	**wy**	**yc**	**wy**	**yc**
0	0.0	0	0	0	0	1	2	0	-2
1	0.0	-1	-4	1	-2	0	-2	0	2
2	0.0	0	0	0	0	-1	2	1	0
3	0.0	0	0	-1	-2	0	-2	0	-2

TABLE 6.3. How **CONV** works for a derivative of a function.

Obviously we can compute the second partial derivatives in the same way, e.g. $\partial^2 / \partial x_1 \partial x_2$ by choosing **sym** $= 3 = 1 * 2^0 + 1 * 2^1 + 0 * 2^2$.

In the example the parameter **or** will be $(1, 1, 1, 1)$ as all bins are present in **xb**. It can happen that we have a nonempty bin which would not appear in **xb** since no observation falls into this bin. **CONV** will generate all bins from min(**xb**) to max(**xb**). If we now compute the kernel density estimate we get an estimation for all bins. This is important for graphical purposes. But if we are only interested in the estimation at X_i as in EPP we need a possibility to identify the original X_i in **xc**. **or** contains a one if **xc**$_i$ is in **xb** or else a zero.

The command `PAF` will kill all columns which do not come from the original X_i.

6.3 Selected Tools in XploRe 3.2

6.3.1 *Multivariate Normal Random Generator*

A multivariate normal random generator can be computed by the macro `NORMGEN`. The syntax is

```
x = NORMGEN(n mu sigma)
```

with `n` a scalar giving the number of observations, `mu` a $p \times 1$ matrix for the mean and `sigma` a $p \times p$ matrix for the covariance matrix.

It will be computed by

$$\mathtt{mu} + \mathtt{sigma}^{-1/2}\mathtt{NORMAL}(\mathtt{n}\,\mathtt{p}).$$

We can compute only one multivariate normal dataset with this command.

6.3.2 *Principal Component Analysis*

In `XploRe` we have two macros for principal component analysis: `PCA` and `HDPCA`. We will concentrate on the macro `HDPCA`. In Figure 6.1 we see the principal component analysis of the Swiss banknote dataset. The display consists of four windows:

- The two windows in the upper rows which show a plot of the first and the second principal component (left) and a plot of the third and the fourth principal component (right). Both plots are linked and we can use the brush to mark the datapoints. If the datapoints have a prespecified colour, e.g. the three clusters marked by the result of EPP, these colours will be used in the plot.
- The left lower window shows the eigenvectors for the principal components. We can see which variables influence a component most. It is possible to scroll through the text to see the hidden components.
- The last window contains the scree plot. The plot shows the explained total variance for each component. The vertical lines tell us how many components will be included by the various criteria (left: elbow, middle: Kaiser, right: 90%). The crosses above and below the stars are the

confidence intervals for $\alpha = 0.05$. Additionally it is tested if the eigenvalues are equal from the last to the first ($\alpha = 0.05$). In this case the corresponding stars will appear in red instead of blue.

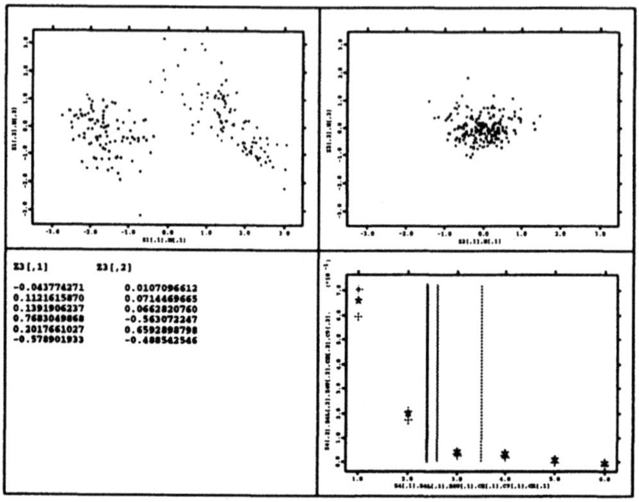

FIGURE 6.1. Principal component analysis of the Swiss banknote dataset.

The syntax of the PCA-macro is

```
y = HDPCA (x)
```

with **x** our data and **y** the principal components with the maskvector which could be changed via brushing.

6.3.3 Grand Tour

The syntax of the grand tour macro is

```
m = TOUR (x)
```

with **x** our multivariate data and **m** a mask vector which could be changed via brushing.

The grand tour macro shows the actual projection (see Figure 6.2). The datapoints are coloured before the call of the grand tour so that the genuine banknotes are marked with blue stars and the forged banknotes with red crosses. In the right window we see the actual projection vectors. We use the subspace interpolation algorithm. Whenever it looks as if the sequence of projections is interrupted, we have reached an endpoint, and a new random

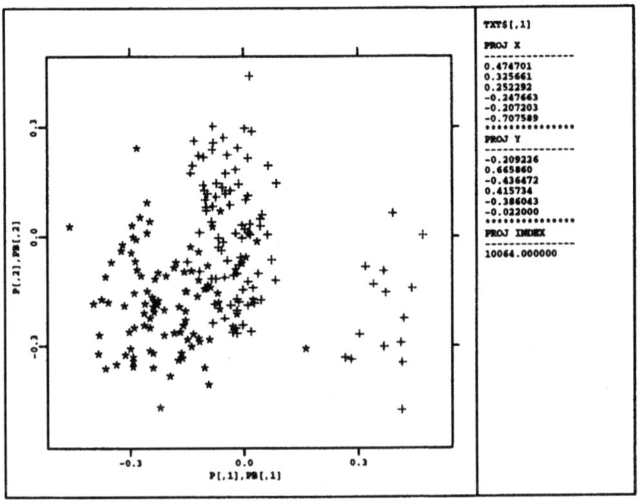

FIGURE 6.2. Grand tour of the Swiss banknote dataset. The picture of the data is manually focussed.

projection will be generated. The same behaviour can be observed in **XGobi** if the number of datapoints becomes large.

The last figure makes the whole projection reconstructable. It is the one which is used for the random seed:

```
randomize(i*pi).
```

The grand tour offers the keypresses

$< + >$ that i for each projection is increased by one

$< - >$ that i for each projection is decreased by one (reversion of the direction)

<Ins> toggles the size of the output. As a standard the picture will be scaled on $[-1, 1]^2$ and will not fit to the data.

<Enter> toggles the moving mode. The standard is not to wait in the picture, but to compute immediately the next projection. If the mode is toggled the program will wait in the picture and we the possibility to brush parts (see for example in Härdle & Simar (1995) Figures 16.3 and 16.4).

6.3.4 Multidimensional Scaling

We have two versions of multidimensional scaling available:

- **MDS**
  which expects a matrix of distances and performs a metric scaling. The call is

  `(e v) = MDS (x)`

  with **x** a distance matrix and **v** the proportion of variance explained by several dimensions. If we consider d dimensions to be sufficient we have to take the first d columns of e to get an approximate picture of the distribution of the datapoints belonging to this distance matrix.

- **NMDS**
  which expects a vector of distances and performs a nonmetric scaling with the stress function S_2.

6.3.5 Exploratory Projection Pursuit

The macros for exploratory projection pursuit are part of the **ADDMOD** library. The following sequence in **XploRe**

```
x = read ("bank2")
library ("addmod")
ch = 0
p = ppexpl (x ch)
```

will load the Swiss banknote dataset and call the macro for exploratory projection pursuit. The data will automatically be centered and sphered. The screen is divided into four windows. The dataset is projected along the first two principal components. On the right side a menu appears.

At first we will be asked whether the output should be coloured or in gray scale. This is only important if we want to plot index functions; see Figures 4.6 - 4.15. The next menu consists of two parts. The **SEARCH PROJ** part tries to find an optimal projection according to the selected index function. The actual best projection is displayed in the 2D-window. The left lower text-window gives informations about the selected index function, the smoothing parameter for the density estimation, the value of the index function for the actual projection, and a step parameter. The **Reset** selection allows us to reset the projection to the initial projection based on principal component analysis. The **Reinit** selection selects a random projection as initial projection. The **INDEX FUNCTION** makes a contour plot of the index function.

The additional parameter **ch** tells the macro that the whole process is interactive if it contains only a zero. If you put in a float vector as **ch**, the macro can run completely noninteractive. The float vector **ch** should have the following entries:

ch[1,1]	=	3	Use colour
		4	Use gray scale
ch[i,1]	=	5	Find best projection with Friedman-Tukey index
		6	Find best projection with entropy index
		7	Find best projection with Legendre index
		8	Find best projection with Hermite index
		9	Find best projection with Natural-Hermite index
ch[i+1,1]	=	Bandwidth or order	
ch[i,1]	=	16	Index function with Friedman-Tukey index
		17	Index function with entropy index
		18	Index function with Legendre index
		19	Index function with Hermite index
		20	Index function with Natural-Hermite index
ch[i+1,1]	=	Bandwidth or order	
ch[i+2,1]	=	Number of gridpoints	
ch[i,1]	=	0	Quit
		1	Quit
		11	Reset
		12	Reinit

The numbers presented here may change as the macro is still under development. The correct numbers can be found by counting on which line of the menu the appropriate entry appears.

Any other value of **ch[i,1]** will have no effect. In the noninteractive mode, the program will give the result to a specified device such as a printer.

For example, the following program

```
proc()=main()
  x = read ("bank2")
  x = x[,4:6]
  library ("addmod")
  ppexpl (x #(3 20 5 50 0))
endp
```

will draw coloured contour lines of the index function for the Natural-Hermite index with order 5 for the last three variables of the Swiss banknote dataset based on a grid with $50 \times 50 = 2500$ projections. The 0 quits the **PPEXPL** macro.

EPP is a technique that allows us to use interactive graphics for interpretation. The macro

```
ppinter (x p)
```

links the projected dataset of the EPP to each of the variables (strips of data in lower plot, see also Figures 4.27 - 4.29). By simple linking we can visualize the relationship between the projection and the original variables. In the upper plot, we see the best projection found with **PPEXPL**. The lower plot shows jittered dotplots of each of the variables. The variables are rescaled on $[0,1]$ via $xr_j = (x_j - \min_j(x_j))/(\max_j(x_j) - \min_j(x_j))$. The text window on the right shows the projection vector for X and Y found by EPP. Additionally you can see the result of the χ^2-test described.

We can brush now in the upper plot and study how the brushed datapoints will appear in the coordinate (variable) axes to assist interpretation.

Another possibility would be to mark each variable to see how it behaves in the projection.

6.3.6 Interactive Clustering

Since the computational power has increased we are now able to do clustering interactively.

As usual it is difficult to represent multivariate data. Some preprocessing like principal component analysis or factor analysis will have to be done . We have chosen to show the plot of the first three principal components (see Figure 3.2). Another possibility would be to use trivariate MDS. It might be helpful to represent the results of the cluster analysis with a weighted principal component plot.

The interactive clustering menu offers the following possibilities:

Picture allows to enter into the 3D-scatterplot. All possibilities of 3D--scatterplot are available (rotating, masking, etc.).

***k*-means** the k-means algorithm.

Adaptive an adaptive k-means algorithm described in Mucha & Klinke (1993).

Ward the Ward algorithm (only euclidean distance).

Single the single linkage algorithm.

Mean link the mean link algorithm.

Median link the median link algorithm (only euclidean distance).

Average the average linkage algorithm.

Centroid the centroid linkage algorithm (only euclidean distance).

Lance the flexible method of Lance and Williams.

Redo repeats the last method, but we are able to use another number of clusters

The distances which are offered for the agglomerative methods are the euclidean distance, the L_1 distance, the L_∞ distance, the cosine distance and the χ^2 distance. For some methods the choice of distances is limited to specific ones. The macro is only developed for the use of continuous variables.

The screen is divided into four pictures: the main picture shows the 3D-scatterplot of the first three principal components. The lower left picture contains the dendrogram and the lower right shows the selected method and the parameters of the method.

Sometimes is it useful to classify on variables instead on observations. This can easily be achieved by transposing the data matrix and then using the cluster algorithm.

6.3.7 Kernel Regression

Macro	Kernel
`UNI`	Uniform product kernel
`TRIAN`	Triangle product kernel
`EPA`	Epanechnikov product kernel
`QUA`	Quartic product kernel
`TRI`	Triweight product kernel
`GAU`	Gaussian kernel
`COSI`	Cosine product kernel
`RUNI`	Radial symmetrical uniform kernel
`RTRIAN`	Radial symmetrical triangle kernel
`REPA`	Radial symmetrical epanechnikov kernel
`RQUA`	Radial symmetrical quartic kernel
`RTRI`	Radial symmetrical triweight kernel

TABLE 6.4. Kernels implemented as macros in `XploRe`

We have various macros available for kernel regression:

- **SKERREG**
  computes the multivariate Nadaraya-Watson estimator without binning. Only the quartic kernel is available. The command SKER is the basis of this macro:

  ```
  (r f) = SKER (x y h xest).
  ```

  It computes the denominator for `xest` and the counter separately. The macro only computes `r/f`.

- **REGEST**
  computes an univariate Nadaraya-Watson estimator with binning. It uses mainly the commands BINDATA and CONV. With BINDATA we bin the data and with the macro SYMWEIGH we produce a binned version of the kernel, here the quartic is taken, but we could also take any other kernel from Table 6.4 or program it by ourselves. The result of SYMWEIGH is that the binned wy-values add up to one.

  Then CONV is used to compute the denominator and the counter separately, and a simple division computes the final estimator.

- **REGAUTO**
  Works like REGEST, but the bandwidth is chosen as $0.1 * (\max(x) - \min(x))$.

- **REGESTP**
  Works like REGEST, but computes a multivariate Nadaraya-Watson estimator.

Although the binned estimator should be faster than the direct estimator the inverse is true. The direct estimator is completely written in C, whereas the binned estimator runs in the macro language which is much slower.

6.3.8 Projection Pursuit Regression

The PPR is part of the ADDMOD-library of XploRe 3.2 (additive modeling). We have two macros available, PPR does the projection pursuit regression, PPRINTER helps to analyze the found projections and fits. The calls are

```
(xs ys vs fs) = PPR (x cmd)
PPRINTER(x vs fs)
```

The PPR-macro allows an interactive fitting of a PPR model with different smoothers. The macro call is:

```
LIBRARY("addmod")          ; Load the addmod library
```

```
x = READ("djppr")             ; Read the dataset generated from
                              ; Donoho and Johnstone (1989)
(xs ys vs fs) = PPR(x)        ; Do the PPR
```

The macro input is a dataset **x**. It is expected that the last column of **x** contains the regressor. The macro output are the standardized data, so **xs** has the property $E(X_i) = 0$ and $Var(X_i) = 1$. The same holds for **ys**. **vs** contains the projection vectors found by a m-term fit and **fs** the fitted function values.

If you enter the macro you will get a menu where you can choose between different items. Since each fit of a m-term model is saved automatically in a PostScriptfile the first item allows you to change the names of the PostScript-files. The standardname is TERM1.PS, TERM2.PS, TERM3.PS and so on. If you change the name to SUPSMO the name of the outputfiles will be SUPSMO1.PS, SUPSMO2.PS, SUPSMO3.PS and so on.

The second item `Cycle Proj` allows to look at random projections $(\alpha^T X_i, Y_i)$ of our data to get an impression. All other choices start the PPR with different smoothers. You will always be asked how many terms you want to fit. The next question asks for the smoothing parameter except in the case of the supersmoother.

You will get a sequence of pictures. In the big window you see the actual fit and the data which are fitted. The window under the big window shows how the error function will increase by changing the projection vector. The dotted line shows the value of the interpretability index of Morton (1989). The right lower window shows how the error in each term decreases while the upper window shows the actual projection.

The second macro **PPRINTER** will help the user to interpret the result. The first item `Enter Pic` allows us to enter the picture of the projections $(Y_i, Y_i - \hat{Y}_{i,m})$ and to use the facilities of the plot. The title tells us how large m is at the moment and how many terms are available (M) at all. The lower right value shows the R^2 value. The second item `Circle Term` cycles from $m = 1, ..., M$ computed by **PPR**. The third term `Set term` allows to choose a fixed m. The fourth and fifth item allow to see the loadings and the squared loadings to analyze which term has most influence to a projection.

In contrast to **S-Plus** **XploRe** allows the user to have a look at the algorithms used since everything is written in macro language. The disadvantage is that **XploRe** needs 15 minutes to fit one term for the Boston Housing Data on a 486DX4-100 PC. To avoid getting bored **XploRe** offers the possibility to work interactively or to start a batch job. A second optional parameter in the call of `PPR (x cmd)` allows the user to give a list of numbers to the macro which encode the different options (see Table 6.5).

cmd[i,1]	= 0	Quit macro
cmd[i,1]	= 1	Change print name to an integer
	cmd[i+1,1]	New name
cmd[i,1]	= 2	Cycle through random projection (endless loop!)
cmd[i,1]	= 4	Use polynomial fitting
	cmd[i+1,1]	Number of terms to fit
	cmd[i+2,1]	Order of polynomial
cmd[i,1]	= 5	Use running median fitting
	cmd[i+1,1]	Number of terms to fit
	cmd[i+2,1]	Number of included neighbours
cmd[i,1]	= 6	Use symmetrized *k*-nearest-neighbour fitting
	cmd[i+1,1]	Number of terms to fit
	cmd[i+2,1]	Number of included neighbours
cmd[i,1]	= 7	Use the supersmoother
	cmd[i+1,1]	Number of terms to fit
cmd[i,1]	= 8	Use the Nadaraya-Watson estimator
	cmd[i+1,1]	Number of terms to fit
	cmd[i+2,1]	Bandwidth
cmd[i,1]	= 9	Use the smoothing spline
	cmd[i+1,1]	Number of terms to fit
	cmd[i+2,1]	Lambda
cmd[i,1]	= 10	Use a haar wavelet estimator
	cmd[i+1,1]	Number of terms to fit
	cmd[i+2,1]	Coefficients used for hard threshold
cmd[i,1]	= 11	Use a local polynomial estimator of order 3
	cmd[i+1,1]	Number of terms to fit
	cmd[i+2,1]	Bandwidth

TABLE 6.5. Command options for a batch job for PPR in `XploRe`

With the PPR-algorithm of `XploRe` and `S-Plus` we get different results for the projection vectors. We do not believe that this due to the missing backfitting in `XploRe`. Since the computed R^2 for the Boston Housing data are practically identical we believe that the error function in dependence of the projection has more than one maximum which seems to have almost the same height. Different choices for the projection vector in early terms lead to different projections fits in later terms.

If we see PPR as a dimension reduction technique we have to base our decision on the corresponding projection of the standardized variables.

Proj	CRIM	ZN	INDUS	CHAS	NOXSQ	RM	AGE	DIS	RAD	TAX	PTRATIO	B	LSTAT
Lin.Reg.	-0.02	+0.00	+0.00	+0.16	-0.01	+0.01	+0.00	-0.33	+0.17	-0.00	-0.05	+0.64	-0.65
XploRe 1	+0.01	+0.00	+0.00	-0.25	+0.01	-0.01	-0.00	+0.31	-0.14	+0.00	+0.10	-0.31	+0.83
XploRe 2	-0.00	+0.00	+0.02	-0.63	-0.00	-0.00	+0.00	-0.64	+0.20	+0.00	-0.01	-0.35	-0.07
XploRe 3	+0.00	-0.00	+0.00	+0.13	-0.00	-0.03	+0.00	+0.28	+0.10	+0.00	-0.01	-0.66	-0.67
XploRe 4	-0.00	-0.00	-0.00	-0.76	+0.00	-0.00	-0.00	-0.13	+0.16	+0.00	-0.01	-0.59	-0.11
XploRe 5	+0.02	-0.00	+0.00	-0.06	-0.01	-0.08	+0.00	-0.05	-0.28	-0.00	-0.08	+0.02	-0.95
XploRe 6	+0.00	+0.00	+0.02	+0.11	-0.00	+0.00	+0.00	+0.07	+0.02	+0.00	+0.02	+0.99	+0.02
XploRe 7	+0.00	-0.00	+0.01	+0.37	+0.00	-0.01	+0.00	+0.05	+0.02	-0.00	-0.04	-0.92	-0.08
XploRe 8	-0.00	-0.01	+0.01	+0.06	-0.01	+0.00	+0.01	+0.41	+0.20	-0.00	+0.04	-0.89	-0.05
S-Plus 1	-0.02	-0.00	+0.00	+0.11	-0.01	+0.01	-0.00	-0.21	+0.22	-0.00	-0.08	+0.44	-0.82
S-Plus 2	+0.00	+0.00	-0.01	-0.01	-0.02	-0.00	+0.00	-0.92	+0.04	-0.00	+0.01	-0.36	-0.01
S-Plus 3	+0.01	+0.00	-0.00	-0.03	+0.00	+0.02	-0.00	-0.42	-0.33	-0.00	+0.02	+0.51	+0.65
S-Plus 4	-0.02	+0.00	-0.01	-0.24	-0.00	+0.04	-0.00	-0.37	+0.30	+0.00	+0.03	-0.34	+0.76
S-Plus 1	-0.02	-0.00	-0.00	+0.11	-0.01	+0.02	+0.00	-0.27	+0.14	-0.00	-0.08	+0.67	-0.66
S-Plus 2	+0.00	+0.00	-0.02	+0.01	-0.02	-0.00	+0.00	-0.97	+0.07	-0.00	+0.02	-0.25	+0.02
S-Plus 3	+0.00	+0.00	-0.01	+0.09	-0.00	+0.02	-0.01	-0.33	-0.20	-0.00	+0.01	+0.70	+0.59
S-Plus 4	-0.02	+0.00	-0.02	-0.28	-0.01	+0.05	+0.00	-0.41	+0.31	+0.00	+0.03	-0.40	+0.70
S-Plus 5	+0.00	-0.02	+0.11	+0.08	+0.02	+0.04	-0.02	+0.82	+0.24	-0.00	+0.02	-0.43	+0.23
S-Plus 6	-0.01	+0.01	-0.01	+0.68	-0.02	-0.02	+0.00	-0.21	-0.60	+0.00	+0.11	+0.29	-0.21
S-Plus 7	+0.03	-0.00	-0.03	-0.38	-0.00	-0.03	+0.00	-0.35	+0.40	-0.00	-0.00	+0.25	-0.71
S-Plus 8	-0.01	-0.00	+0.15	+0.42	-0.08	-0.01	+0.01	-0.61	+0.13	+0.00	-0.17	-0.12	-0.60

TABLE 6.6. Projection vectors from PPR and the linear regression for the Boston Housing Data. The coefficients of the linear regression (first line) are rescaled so that the euclidean norm is 1. Then we see the projections of 8-term-PPR-fit in `XploRe` and the projection of a 4-term- and 8-term-fit in `S-Plus`.

6.3.9 Wavelet Regression

For the wavelet estimation we have several commands available:

- **WAVEGEN**
  which allows to generate the father and mother wavelet from the wavelet constants h_k and some other parameters. The wavelets constants are available for the Daubechies wavelets (Haar/D2, D4, D6, ..., D20), the Symmlets (S4, S5, ..., S10) and the Coiflets (C1, C2, ..., C5), and they are stored in the file **DATA\XWAVELET.DAT**.

- **WAVEEST**
  estimates the coefficients for the father and the mother wavelets. The procedure has an optional parameter **y** if a regression is to be performed, otherwise a wavelet density estimation is performed.

- **WAVESMO**
  constructs the estimation curve from the possibly thresholded coefficients.

- **FWT** and **INVFWT**
  are contributed by Prof. Golubev, Moscow. This algorithm assumes periodic wavelets and uses the cascade algorithm.

Since the function **WAVEEST** does not use the cascade algorithm we can compute local thresholds based on:

$$t_{jk} = C\sqrt{2\sigma_{jk}^2 \log(\#\text{ thresholded constants })}$$

with

$$\hat{\sigma}_{jk}^2 = \frac{1}{n}\sum_{i=1}^{n} \varphi_{jk}^2(X_i)\left(Y_i - \frac{Y_{i-1}+Y_{i+1}}{2}\right)^2.$$

The macro **WAVELET1** is built upon the three first commands. The syntax is

WAVELET1(x)

with **x** a two dimensional nonequidistant data matrix. The first step is to bin in a way that we get an equidistant design. Then a regression estimation is computed for the Haar wavelets with setting all coefficients for the mother wavelets to zero. The result is displayed. The display used consists of four windows. The upper left is used to display the data and the fit, in the upper right the menu appears. The lower left contains the coefficient of the mother wavelets, all coefficients not used appear in blue. If a coefficient is used in an estimate it will appear in red. We see the coefficient close to the place where it will have influence (locality of the wavelet) on the estimate. Each row represents a layer.

We have now the possibility to brush the coefficients with red and blue. If we leave the macro with `<Esc>` the program will compute the new fit accordingly to the brushed coefficients. With `<F3>` we ask for the true values and variances of a specific coefficient.

If we do not brush any coefficient we enter the menu with the following items:

- **`Choose basis`**
  allows us to choose another basis than the Haar wavelets

- **`Choose n`**
  allows us to choose more datapoints for binning

- **`Choose level`**
  allows us to choose other layers for estimating, the default is lowest layer $\sim log_2(n)/5$ and highest layer $\sim log_2(n/\log(n))$

- **`Toggle data`**
  shows the binned data in the plot instead of the true data

- **`Hard global`**
  applies a hard threshold t to the mother wavelet coefficients. The absolute value of the coefficients will be shown in the right lower window

- **`Soft global`**
  applies a soft threshold t to the mother wavelet coefficients. The absolute value of the coefficients will be shown in the right lower window

- **`Hard local`**
  applies a hard threshold t_{jk} to the mother wavelet coefficients. A transformation of the coefficients divided by the estimated variances will be shown in the right lower window

- **`Soft local`**
  applies a soft threshold t_{jk} to the mother wavelet coefficients. A transformation of the coefficients divided by the estimated variances will be shown in the right lower window

- **`And`**
  applies the following brushing- or threshold operations only to the coefficients actually used

- **`Copy`**
  applies the following brushing- or threshold operations to all coefficients

6.3.10 Interactive Contouring

As already shortly described **XploRe** is able to generate contour plots. The macro **contour** of the **HIGHDIM** library provides an easy tool for interactive contouring for 2D- and 3D-surfaces. The only assumption which is needed is that the data are provided in a special form. The data have to be on the vertices of a rectangular mesh. So our dataset has the form

$$(x_i, y_j, f_{ij}) \quad \text{or} \quad (x_i, y_j, z_k, f_{ijk})$$

with $i = 1, ..., n_i, j = 1, ..., n_j, k = 1, ..., n_k$. Let us as an example compute the density of three variables of the Swiss banknote data: the width of the banknote, the height of the left side, and the diagonal of the inner box. First we load the libraries **HIGHDIM** and **SMOOTHER**, then we construct our dataset:

```
library ("highdim", "smoother")
x = read ("bank2")
x = x[,1]~x[,2]~x[,6]
```

Since we want to make a kernel density estimate at the grid points we compute the minima and the gridwidth in each variable:

```
x0 = min(x)
gw = (max(x)-min(x))./15
```

Then we generate a grid. For the kernel estimate we need additionally a bandwidth h, then we can easily compute a kernel density estimate for each gridpoint $(15 + 1 = 16)$.

```
nh = 5
h  = nh.*gw
(xb yb) = bindata (x gw)
wx = matrix(cols(x) 1 0)
wy = symweigh (wx 1./nh nh &qua)
wx = grid(wx 1 nh)
(xc yc or) = conv (xb yb wx wy)
xs = (xc.*trn(gw))~(yc./(rows(x).*prod(h)))
xs
```

xs now contains the gridpoints and a density estimate. Looking at **xs** you will see how the order of the gridpoints should be like in order to pass them directly into the **contour** macro. To generate a 2D-surface for a two dimensional dataset we simply have to generate an **x** with two columns. For a 2D-surface we may increase the number of gridpoints from 16 to 50. Please keep in mind that for a 3D-surface we have generated $16^3 = 2048$ gridpoints and for a

2D-surface we would have $50^2 = 2500$ gridpoints. To choose 50 gridpoints for each variable for a 3D-surface would result in $50^3 = 125000$ gridpoints. Depending on our installation this may cause a fatal error in **XploRe** because of memory problems; and surely the execution of the program would take a long time.

Now we call the contouring macro with our dataset **xs**. Since we may want to store some contour plots in files we use the **capture** command. With the parameter **on**, we force **XploRe** to ask us before printing in which file we want the picture to be saved. If we would use another parameter here, **mycont.ps** for example, the parameter would be used as a name for the graphic file. But if we want to store more than one picture the second picture will overwrite the first, the third the second and so on.

```
capture ("on")
contour (xs)
```

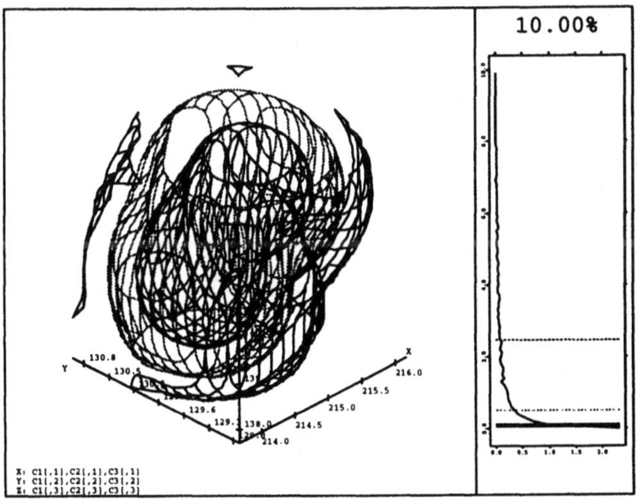

FIGURE 6.3. Kernel density estimate of three variables of the Swiss banknote dataset; automatic chosen contours.

The result of these commands is shown in Figure 6.3. The right window of the plot shows us some informations about our contour lines and our densities. We have computed a density estimate f_{ijk} for each gridpoint (x_i, y_j, z_k); it is an easy task to rescale the f_{ijk} so that

$$\tilde{f}_{ijk} = \frac{f_{ijk} - \min_{i,j,k} f_{ijk}}{\max_{i,j,k} f_{ijk} - \min_{i,j,k} f_{ijk}}.$$

It holds that $\tilde{f}_{ijk} \in [0, 1]$. Which contours are now plotted? You see a blue, green, and red line in the right window which corresponds to values near 0.0, 0.05, and 0.25. So we have plotted the contours for

$$\begin{array}{rcll} \tilde{f}(x, y, z) & \sim & 0.00 & \text{blue, outer} \\ & \sim & 0.05 & \text{green, middle} \\ & \sim & 0.25 & \text{red, inner.} \end{array}$$

The white (in the graphic black) line gives us informations on how many $\tilde{f}_{ijk}$ we have at a specific contouring level. Although we have chosen a big bandwidth we can see that most of the generated gridpoints have a density of 0, which is the smallest value we can get. The "curse of dimensionality" is present here!

Before we start to change the contour levels, let us have a look how **XploRe** chooses the contour levels. **XploRe** takes the 20%, the 50%, and the 80% quantile of the $\tilde{f}_{ijk}$. The blue line represents the 20% quantile, the green line the median, and the red line the 80% quantile. It may happen that they coincide, so, as an example, make the bandwidth smaller (**nh = 2**), and you will miss one line, but **XploRe** will always compute three contour levels.

Now use the keys <↑> and <↓> to change the contour level of the blue line. This is the active line and it is plotted as a thick solid line, whereas the others are plotted as thin dashed lines. With <←> and <→> we can change how much <↑> and <↓> will increase our active level. The current value is plotted in the right window at the top, it is 10%.

The keys **<Pg Up>** and **<Pg Dn>** activate one of the other lines. Another picture which you can obtain by changing the contour levels is shown in Figure 6.4.

Maybe you will not be satisfied with the actual perspective you have from the contour plot. Use **<ENTER>** to enter the contour plot. Since this a normal 3D-window of **XploRe**, you can use all facilities of a 3D-picture, like rotating, brushing, and zooming. Leave the plot with **<ESC>** and the contouring macro with another **<ESC>**.

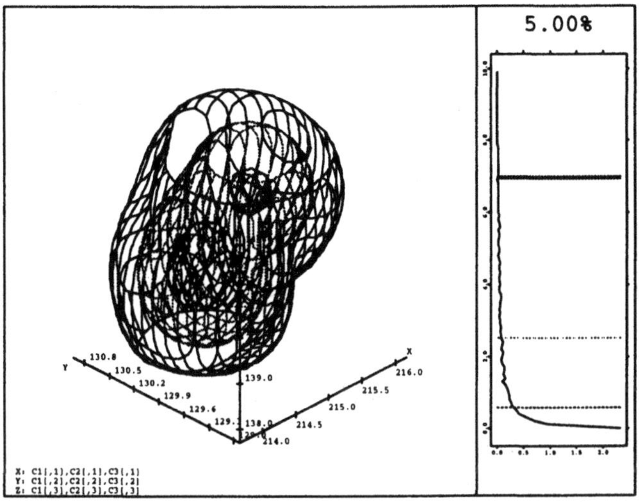

FIGURE 6.4. Kernel density estimate of three variables of the Swiss banknote dataset; user chosen contours.

6.4 Data Structure in XploRe 4.0

6.4.1 Numerical Arrays

As explained in section 5.2.1 we need multidimensional arrays to store our data. In **XploRe** 3.2 we only have matrices like in many other statistical programs.

The basic object in **XploRe** 4.0 is a eight dimensional array. We only need to change the value of the constant **MAX_XARRAY** from 8 to another figure to decrease or to increase the dimensionality of the arrays. The programming of the arrays in C++ allows us to build up a hierarchy of classes and to reduce the amount of programming.

Class hierarchy is:

- **XplArray** — basis array
  - **XplNumber** — basic number
    - **XplInteger** — basic integer
      - **XplShort** — short
      - **XplUshort** — short
      - **XplInt** — int

- · **XplUnsigned** unsigned
- · **XplLong** long
- · **XplUlong** unsigned long
- * **XplReal** basic float number
  - · **XplFloat** 4-byte float
  - · **XplDouble** 8-byte float
  - · **XplLdouble** 10-byte float

The basis class **XplArray** contains operations which are the same for all necessary datatypes like

- Construction and destruction of arrays
- Creating of subarrays
- Indexing of arrays
- Concatenation of arrays
- Filling of arrays
- Reshaping of arrays
- Assignment of arrays
- Sorting of arrays
- Sizes of arrays
- Test if matrices are conformable
- Casting
- Reading and writing of arrays (screen/files) in ASCII or X-format
- Layer, matrix and vector dummy procedures

Step by step we are specializing the classes. **XplNumber** contains all basic operations which can be done on numbers (here float numbers and integers) like

- Elementwise unary mathematical operations (unary plus, unary minus)
- Elementwise binary mathematical operations (addition, subtraction, multiplication, division, exponentiation)
- Assignment

- Transposing of matrices
- Multiplication of matrices $(X * Y, X^T * Y)$
- Multivariate k-nearest-neighbour
- Binning
- Convolution
- Constructing grids
- Conditional sums
- Conditional products
- Conditional minima
- Conditional maxima
- Conditional cumulative sums
- Conditional cumulative products.

Now we split them into two classes: one for float numbers (`XplReal`) and one for integers (`XplInteger`). The `XplReal` class incorporates the following operations

- Elementwise mathematical functions (e.g. cosine, sine, etc.)
- Computing the inverse of a matrix
- Random generators
- Fast fourier transform
- Eigenvector and eigenvalue calculation for symmetric matrices.
- Conditional means
- Conditional variances.

A finer class is obtained by deriving classes for different levels of precision of computing: `XplLdouble` for 10-byte float numbers, `XplDouble` for 8-byte float numbers and `XplFloat` for 4-byte float numbers.

As mentioned earlier the amount of programming code is reduced by using derived classes which is possible by the ability of *inheritance* (defining an operation in `XplNumber`), with the result that we have the operation available in all derived classes (`XplReal`, `XplFloat`, ...).

Obviously we need another feature of C++ called *Template*, which means we do not fix the datatype (e.g. float, double, ...).

The **XplInteger** class supports

- Elementwise logical operations (AND, OR).

Special classes are also derived for different type of integers: signed integer (the size, 4- or 2-byte, machine dependend), unsigned integer, signed long (4-byte integer) and unsigned long.

6.4.2 String Arrays

Another class of arrays has to handle strings. A problem that frequently occurs with data is that we have nominal variables which are represented as text. In **XploRe** 3.2 we were forced to change the **READ** command as we can handle 16384 strings at maximum. So if we read a big dataset like the Berlin Housing Data where we have 3 yes/no variables (FM, FB, FE) we would have 45000 strings to store. This is the reason to create a class **XStringDatabase** as a skip list (Schneider 1994) such that the same string is only stored once, but referenced more often. In the case of the Berlin flat data this would result in 45000 references and only two entries in the skip list: "yes" and "no".

- **XStringDatabase** — database for strings
- **xstring** — handling of strings
- **XplArray** — basis array
  - **XplChar** — basic text

The class **xstring** offers the standard string operations and the class **XplChar** handles arrays of strings.

6.4.3 Other Arrays and their Implementation

We need some additional classes of arrays, e.g. **XplColour** derived from **XplArray** to handle arrays of colours.

The class **XplIndex** handles the multidimensional indexing of the arrays.

This represents the actual state, since **XploRe** 4.0 is still under development.

6.5 Commands and Macros in XploRe 4.0

6.5.1 BINDATA *and* CONV

The syntax for **BINDATA** and **CONV** is almost the same as in **XploRe** 3.2:

```
(xb, yb) = BINDATA(x, binwidth, origin, y)
(xc, yc, or)=CONV(xb, yb, wx, wy, sym).
```

What has changed is that we have the possibility to write

```
bdata = BINDATA(x, binwidth, origin, y)
cdata = CONV(xb, yb, wx, wy, sym)
```

so that **bdata** has subobjects which can be accessed via **bdata.xb**, **bdata.yb** and in the case of **CONV** via **cdata.xc**, **cdata.yc** and **cdata.or**.

As mentioned earlier we have multidimensional arrays as input and output. The sizes are given in Table 6.7 for **BINDATA** and in Table 6.8 for **CONV**.

Parameter	Size
x	$n \times p \times q_1 \times \ldots \times q_k$
binwidth	$p \times 1 \times r_1 \times \ldots \times r_k$
origin	$p \times 1 \times s_1 \times \ldots \times s_k$
y	$n \times m \times t_1 \times \ldots \times t_k$
xb	$n \times p \times v_1 \times \ldots \times v_k$
yb	$n \times m \times v_1 \times \ldots \times v_k$
	with $v_l = \max(q_l, r_l, s_l, t_l)$

TABLE 6.7. Sizes of the input and the output parameters of **BINDATA**. For each l only one the parameters q_l, r_l, s_l and t_l can be different from 1.

We note that **BINDATA** does no longer shorten the output array **xb** and **yb** in the size as in **XploRe** 3.2. This due to the fact that we allow different binwidths and origins in the higher dimensions. If we do so the number of nonempty bins may vary, but we are sure it can be larger than n. Thus we summarize up as in **xb** and **yb**, but we put also the nonempty bins in. The entry in **yb** is simply a zero, in **xb** we will get the binnumber of the appropriate observation. We have two possibilities to proceed, we can kill the nonempty bins layerwise or we can compose **xb** with the original **y** to continue with other computations. Since the interplay between **CONV** and **BINDATA** is important **CONV** will be able to react to these results.

The resultsize of **CONV** changes too. If we have different **xb**, **yb** and **wx**, **wy** we will have different grids for the resulting computation, so we compute the largest grid and center the other grids if necessary. We do the computation

Parameter	Size
`xb`	$n \times p \times q_1 \times ... \times q_k$
`yb`	$n \times m \times r_1 \times ... \times r_k$
`wx`	$h \times p \times s_1 \times ... \times s_k$
`wy`	$h \times 1 \times t_1 \times ... \times t_k$
`sym`	$1 \times 1 \times u_1 \times ... \times u_k$
`xb`	$g \times p \times v_1 \times ... \times v_k$
`yb`	$g \times m \times v_1 \times ... \times v_k$
`or`	$g \times 1 \times v_1 \times ... \times v_k$
	with $v_l = \max(q_l, r_l, s_l, t_l)$

TABLE 6.8. Sizes of the input and the output parameters of **CONV**. For each l only one the parameters q_l, r_l, s_l and t_l can be different from 1.

on this large grid.

The size of **sym** is $1 \times 1 \times u_1 \times ....$ We could have started already in the first dimension with a size larger than one. This is not done because of the compatibility to **XploRe** 3.2.

6.5.2 Random Generators

The command for the random generator in **XploRe** 4.0 is

```
x = RANDOM(type, size)
```

with **type** a parameter which distinguishes between different generators, and **size** a 8×1 vector which gives the number of the entries in each dimension.

7 Conclusion

We have presented two versions of the software `XploRe` which implement our ideas about data structures in statistical software. The data structures for graphics and linking are implemented in `XploRe` 3.2 while the data structures for statistical data are implemented in `XploRe` 4.0. But we have not implemented all ideas.

In the chapter 2 we have concluded that we mainly need two types of windows. Some graphics, e.g. the histogram, the contour plot, the Andrews curves and the trees are implemented in `XploRe` 3.2 in the three-step-method as described in the beginning of chapter 5. Other graphics like the Chernoff faces and the boxplot have their own window in `XploRe` 3.2 as proposed at the end of chapter 2 and in chapter 6. The hierarchical graphical structure (datapart, window, display) developed in chapter 5 is implemented in both versions of `XploRe`. A detailed description can be found in the beginning of chapter 6.

Although we have developed different kinds of linkings and have shown the usefulness in chapter 5 (links of plots, of scales and of different data), we have only implemented linking from dataparts to other dataparts or external matrices in `XploRe` 3.2. This already allows to make a lot of interesting and informative links, e.g. for asking datapoints, linking between different plots and so on.

The data structure required for statistical data developed in chapter 5 is only implemented in `XploRe` 4.0. The examples (exploratory projection pursuit, local polynomial regression) for the need of multidimensional arrays can not be described at the end of chapter 6 as we are still implementing the basic graphical data structures and the programming language. So we have restricted ourselves to showing how two commands differ depending on whether we use matrices or arrays. The implementation of hierarchical objects is not easy, and even simple routines like reading and writing the objects produce difficult problems. Some routines can be extended easily, e.g. unary or binary operations, but commands require a rethinking of the whole parameter list.

Further research and work. Although we believe that a statistical software program is composed of three elements, we have only concentrated shortly on the GUI. This is partly due to the fact that `XploRe` 3.2 runs under DOS which has no graphical user interface available from the operating system. This makes necessary an immense amount of additional programming, but this gives us the freedom to develop independently from such systems. This is one of the main reasons which lead to the idea of displays, so GUIs would limit our needs as they do not offer them. Nevertheless for a (economically) successful statistical environment it is necessary to run under a GUI. Even `XploRe` 4.0 now runs under Motif and it is going to run under Windows as well.

With `XploRe` 3.2 we are even able to implement many tools which to me seem to be superior to tools offered by other programs. We provide an interface for linking, interactively and by a command, which can be extended for linking events. Nevertheless I am aware of the fact that the question of linking has not been solved until now. Linking of datasets is only done in `DataDesk`, but it is easy to handle for the user.

Further research for data structures for graphics. In my opinion the research concerning graphical data structures has finished the hierarchy of displays, windows and dataparts. Of course still some detailed work is left concerning the communication between user and graphics (via slider, menus, dialogboxes etc). New developments like window systems and parallelization require very detailed and careful research. Since `XploRe` runs as run-alone program under DOS such considerations have no influence for the development of `XploRe` 3.2. But already the implementation of user-programmable menus in `XploRe` 4.0 causes some serious problems.

Further research for data structures for data. With the introduction of arrays in statistical software the importance of the basic data types decreased. A lot of work has to be done to make arrays behave in a proper way so that functions react to the data, e.g. how will the deletion of an element affect a multidimensional array? The research about hierarchical data objects also needs to be extended. Linking of different data objects still causes serious problems.

Further research for data structures for linking. Up to now linking is always hidden in the programs, e.g. `DataDesk`, only `XploRe` and `X-Lisp-Stat` make it available to the user to some extent. In `XploRe` the linking is connected to the dataparts of a graphic. A first step would be to extend linking on other objects such as the scale of a plot. Linking between different data objects is another problem. The (graphical) visualization of links and hierarchical objects has to be developed as well, but at least we have some ideas how to do it.

Final remark. We hope that this thesis can be a basis for further software development and research in statistical computing and that it will help to simplify the work of software developers. Quite certainly there is a further need for people with a view for both statistics and computer science.

A
The Datasets

A.1 Boston Housing Data

Harrison & Rubinfeld (1978) collected 14 variables (see Table A.1) for 506 districts in the Boston standard metropolitan statistic area at a fixed point of time.

Symbol	Definition
LMV	logarithm of the median value of owner-occupied homes
CRIM	per capita rate by town
ZN	proportion of a town's residential land zoned for lots greater than 25000 square feet
INDUS	proportion of nonretail business acres per town
CHAS	Charles river dummy variable with value 1 if tract bounds on the Charles river
NOXSQ	nitrogen oxide concentration (parts per hundred millions) squared
RM	average number of rooms squared
AGE	proportion of owner-occupied units built prior to 1940
DIS	logarithm of the weighted distances to five employment centers in the Boston region
RAD	logarithm of index of accessibility to radial highways
TAX	full-value property-tax rate (per 10000 US-$)
PTRATIO	pupil-teacher ratio
B	$(Bk - 0.63)^2$ where Bk is the proportion of blacks in the population
LSTAT	logarithm of the proportion of the population of lower status

TABLE A.1. Variables of the Boston housing dataset

They made a linear regression to estimate the relationship to between LMV and the other variables:

$$\begin{aligned} LMV &= \beta_0 + \beta_1 CRIM + \beta_2 ZN + \beta_3 INDUS + \beta_4 CHAS \\ &\quad + \beta_5 NOXSQ + \beta_6 RM + \beta_7 AGE + \beta_8 DIS + \beta_9 RAD \end{aligned}$$

$$+\beta_{10}TAX + \beta_{11}PTRATIO + \beta_{12}B + \beta_{13}LSTAT$$

and obtained the coefficients in Table A.2. The R^2-value is 0.806.

i	Variable	$\hat{\beta}_i$	Standard error	t-statistics
0	CONST	+9.76E+0	1.50E-1	65.23
1	CRIM	-1.19E-2	1.24E-3	-9.53
2	ZN	+7.94E-5	5.06E-4	0.16
3	INDUS	+2.36E-4	2.36E-3	0.10
4	CHAS	+9.14E-2	3.32E-2	2.75
5	NOXSQ	-6.39E-3	1.13E-3	-5.64
6	RM	+6.33E-3	1.31E-3	4.82
7	AGE	+8.86E-4	5.26E-4	0.17
8	DIS	-1.91E-1	3.34E-2	-5.73
9	RAD	+9.57E-2	1.91E-2	5.00
10	TAX	-4.20E-4	1.23E-4	-3.42
11	PTRATIO	-3.11E-2	5.01E-3	-6.21
12	B	+3.64E-1	1.03E-1	3.53
13	LSTAT	-3.71E-1	2.50E-2	-14.83

TABLE A.2. The fitted coefficients of the linear regression. Taken from Belsley, Kuh & Welsch (1980).

A.2 Berlin Housing Data and Berlin Flat Data

The data are collected by Germer & Neudeck (1994) in their master thesis. The dataset consists of two different kinds of data:

- data about offers of houses and
- data about offers of flats ($n = 14968$) (Berlin flat data).

We will restrict ourselves to the last dataset. Table A.3 gives an overview about the collected variables.

We can see from the Table A.3 that we have 3 groups of variables :

- time-dependend:
  The variables T and TI are depended from the time of the offer.

No.	Abbreviation	Meaning
1	T	Time of offer
2	FA	Area of the flat (in m^2)
3	FL	Location of the flat (coding see Table E.1)
4	FR	Number of rooms
5	FB	Existence of a balcony (1=yes, 0=no)
6	FM	Maisonette (1=yes, 0=no)
7	FP	Price in offer (in 1000 DM)
8	FI	Corrected Price in offer (in 1000 DM)
9	FE	Is in the west (1=yes, 0=no)
10	DI	Inhabitant in the district (per ha)
11	DF	Rate of foreigners (in %)
12	DU	Rate of unemployed (in %)
13	DR	Recreation areas, e.g. parks (in %)
14	DW	Blue-collar worker (per ha)
15	DS	Length of public streets (in km per square km)
16	DN	NO_x concentration (in μgram per m^3)
17	DB	Distance to the Breitscheidplatz (in km)
18	TI	Interest rate of the German Bundesbank (in %)

TABLE A.3. Variables of Berlin flat data.

- flat-dependend:
  The variables FA, FL, FR, FB, FM, FP, FI and FE are describing properties of the flat.
- location-dependend:
  The variables DI, DF, DU, DR, DW, DS, DN and DB are describing properties of the location of the flat.

We have a principal problem to analyze the dataset with the aim to model the price of a flat in dependence of the variables. If we only have offers then we do not know for which price a flat has changed its owner. Another problem is that the same flat might have been offered twice at the same or another price. It is impossible to distinguish between a flat which is offered twice and a similar flat on another floor of the same building.

Let us have a closer look at some of the variables:

T The offers are collected from the newspaper "Berliner Morgenpost" which is the most important one to find a flat or a house for renting or buying. Every three months (January, April, July and October) from July 1989 until October 1994 the offers of the first weekend issue are taken.

TI The interest rate of the German Bundesbank is the percentage which other banks have to pay if they borrow money from the German Bundesbank. This influences the percentage someone has to pay if he borrows money from a bank for buying a flat. We will see soon that we have a one-to-one relationship to the variable T.

FL Although the coding of the variable from Table E.1 seems to be clear, we have some problem with it. As the authors of the master thesis remarked by themselves they had offers with inexact locations, like "city" or "at a central place". These offers are coded as number 16 which is the coding number of Charlottenburg, a central district in the west part of Berlin. Contrary offers like "in the north of Berlin" or "in the east part" are dropped from the data.

For example the districts 16 and 30 are very heterogeneous in the prices for offers. Both contain very expensive areas, e.g. like KaDeWe, Kurfürstendamm or Dahlem, and areas of much lower prices, e.g. like Siemensstadt or Lichterfelde.

In their master thesis the authors show a map with a table which connects the coding number to a specific area in Berlin. But on page 47/48 a table is found which also gives a coding number for selected districts and streets. We have to state some contradictions between the map and the table.

FR The number of rooms is not given exactly. We will see later that we have many offers with an integer number of rooms and some offers of half rooms. The number of the latter is much smaller than the other. The authors themselves state that they also have a lot of offers of small flats where no number of rooms is given.

FB The nonexistence of a balcony is questionable. It could be that a balcony is simply not mentioned.

FM Maisonette flats are penthouses, flats in the loft etc. It is expected that higher prices are obtained for that kind of flats.

FI The price of the offer is divided by an index ("Lebenshaltungsindex 1985") which describes the increase of general living costs in the west part of Berlin. This index also includes expenses for clothes, food, public transportation etc. We believe that this index is not appropriate to construct a comparable price over time. The graphics on page 89 in the master thesis show the 25%, 50%, 75% quantiles, the mean value of the prices and the corrected prices over time. We do not see a big difference in the behaviour of these four curves.

A general problem occurs with the variables describing the districts. We will see later that we have only very few different values although we have 44 districts (some of them are even subdivided in two subdistricts; one in the west part and one in the east part).

For the following explanations and examinations we will use the whole dataset or the subset of the offers from October 1994. October 1994 was chosen, because in the newspaper "Der Tagesspiegel" of March/April 1995 we find some data about the living quality in the districts of Berlin. For a discussion of linked matrices see section 5.3.4.

A.3 Swiss Banknote Data

The Swiss banknote data can be found in Flury & Riedwyl (1981). The dataset consists of six continuous variables; see Table A.4. It contains 100 genuine and 100 forged bank notes. The authors believe that they can find five different clusters in the dataset.

L	Length of the bill
HL	Height of the bill on the left
HR	Height of the bill on the right
FL	Distance from the inner frame to the lower bound
FU	Distance from the inner frame to the upper bound
D	Length of the diagonal of the inner frame

TABLE A.4. Variables of the Swiss banknote dataset.

A.4 Other Data

The following other datasets have been used:

Motorcyle The x-values denote time in milliseconds after a simulated impact with motorcycles. The response variable y is the head acceleration in g of a post mortem human test object (Schmidt, Mattern & F. 1981).

RANDU is a 3-dimensional dataset generated from a linear congruential random generator.

TET is a 3-dimensional dataset where the data are distributed on the facettes of a tetrahedron.

CYL is a 3-dimensional dataset where the data are distributed on the surface of a cylinder.

Uniform is an uniform distribution on $[0, 1]^2$.

Normal is a standard normal distributed dataset.

Line is a dataset distributed around a line from $(0, 0)$ to $(1, 1)$ with constant variance.

Two mass is a mixture of two standard normal distributions.

Circle is a dataset distributed with a constant variance around a circle.

B

Mean Squared Error of the Friedman-Tukey Index

Under the assumption that X_i is independent and identical distributed (i.i.d), that the unknown density is sufficiently often differentiable and the following definitions

$$\begin{aligned}
K_{i,j}^{(l)} &= \int_{\mathbb{R}^2} s_1^i s_2^j K(s_1, s_2)^l ds \\
&= \begin{cases} \text{nonzero} & \text{if } i \text{ and } j \text{ even,} \\ 0 & \text{otherwise.} \end{cases} \\
I_j^{(i)} &= \int_{\mathbb{R}^2} f(x_1, x_2)^i \left(\frac{\partial^j f}{\partial x_1^j}(x_1, x_2) + \frac{\partial^j f}{\partial x_2^j}(x_1, x_2) \right) dx_1 dx_2 \\
J &= \int_{\mathbb{R}^2} f(x_1, x_2) f_{x_1 x_1 x_2 x_2}(x_1, x_2) dx_1 dx_2 \\
M_i &= K_{4,0}^{(i)} I_4^{(1)} + 6 K_{2,2}^{(i)} J
\end{aligned}$$

we can start to calculate the MSE of the Friedman-Tukey index. First we transform the Friedman-Tukey index by

$$\begin{aligned}
\hat{I}_h &= \frac{1}{n^2 h^2} \sum_{i=1} \sum_{j=1}^{n} K\left(\frac{X_i - X_j}{h}\right) \\
&= \frac{1}{n^2 h^2} \left(nK(0,0) + 2 \sum_{\substack{i,j=1 \\ i<j}}^{n} K\left(\frac{X_i - X_j}{h}\right) \right),
\end{aligned}$$

which simplifies our calculations.

We can rewrite the MSE

$$\begin{aligned}
MSE(\hat{I}_h) &= E\left((\hat{I}_h - I)^2\right) \\
&= E(\hat{I}_h - I)^2 + Var(\hat{I}_h - I) \\
&= Bias(\hat{I}_h)^2 + Var(\hat{I}_h)
\end{aligned}$$

and the Bias

$$\begin{aligned} Bias(\hat{I}_h) &= E(\hat{I}_h - I) \\ &= E(\hat{I}_h) - I. \end{aligned}$$

The expectation of the estimator is

$$\begin{aligned}
E(\hat{I}_h) &= E\left(\frac{1}{n^2h^2}\left(nK(0,0) + 2\sum_{\substack{i,j=1\\i<j}}^{n} K_h\left(\frac{X_i - X_j}{h}\right)\right)\right) \\
&= \frac{1}{n^2h^2}\left(nK(0,0) + 2\frac{n(n-1)}{2}E\left(K\left(\frac{X-Y}{h}\right)\right)\right) \\
&= \frac{1}{n^2h^2}\left(nK(0,0) + n(n-1)\int_{\mathbb{R}^2} f(x_1,x_2)\right. \\
&\quad \left.\left(\int_{\mathbb{R}^2} K\left(\frac{x_1-y_1}{h},\frac{x_2-y_2}{h}\right) f(y_1,y_2)dy\right)dx\right) \\
&= \frac{1}{n^2h^2}\left(nK(0,0) + n(n-1)\int_{\mathbb{R}^2} f(x_1,x_2)\right. \\
&\quad \left.\left(h^2\int_{\mathbb{R}^2} K(s_1,s_2)f(x_1+hs_1,x_2+hs_2)ds\right)dx\right) \\
&= \frac{K(0,0)}{nh^2} + \frac{n-1}{n}\int_{\mathbb{R}^2} f(x_1,x_2) \\
&\quad \left(\int_{\mathbb{R}^2} K(s_1,s_2)\left(f(x_1,x_2) - hs_1f_{x_1}(x_1,x_2) - hs_2f_{x_2}(x_1,x_2)\right.\right. \\
&\quad +\frac{h^2}{2}s_1^2f_{x_1x_1}(x_1,x_2) + 2\frac{h^2}{2}s_1s_2f_{x_1x_2}(x_1,x_2) \\
&\quad +\frac{h^2}{2}s_2^2f_{x_2x_2}(x_1,x_2) - \frac{h^3}{6}s_1^3f_{x_1x_1x_1}(x_1,x_2) \\
&\quad -3\frac{h^3}{6}s_1^2s_2f_{x_1x_1x_2}(x_1,x_2) - 3\frac{h^3}{6}s_1s_2^2f_{x_1x_2x_2}(x_1,x_2) \\
&\quad -\frac{h^3}{6}s_2^3f_{x_2x_2x_2}(x_1,x_2) + \frac{h^4}{24}s_1^4f_{x_1x_1x_1x_1}(x_1,x_2) \\
&\quad +4\frac{h^4}{24}s_1^3s_2f_{x_1x_1x_1x_2}(x_1,x_2) + 6\frac{h^4}{24}s_1^2s_2^2f_{x_1x_1x_2x_2}(x_1,x_2) \\
&\quad +4\frac{h^4}{24}s_1s_2^3f_{x_1x_2x_2x_2}(x_1,x_2) + \frac{h^4}{24}s_2^4f_{x_2x_2x_2x_2}(x_1,x_2) \\
&\quad -\frac{h^5}{120}s_1^5f_{x_1x_1x_1x_1x_1}(x_1,x_2) - 5\frac{h^5}{120}s_1^4s_2f_{x_1x_1x_1x_1x_2}(x_1,x_2) \\
&\quad -10\frac{h^5}{120}s_1^3s_2^2f_{x_1x_1x_1x_2x_2}(x_1,x_2) - 10\frac{h^5}{120}s_1^2s_2^3f_{x_1x_1x_2x_2x_2}(x_1,x_2)
\end{aligned}$$

$$-5\frac{h^5}{120}s_1 s_2^4 f_{x_1x_2x_2x_2x_2}(x_1,x_2) - \frac{h^5}{120}s_2^5 f_{x_1x_2x_2x_2x_2x_2}(x_1,x_2)$$
$$+O(h^6)\big)\,ds\big)\,dx$$

$$= \frac{K(0,0)}{nh^2} + \frac{n-1}{n}\int_{\mathbb{R}^2} f(x_1,x_2)\left(f(x_1,x_2)\underbrace{\int_{\mathbb{R}^2} K(s_1,s_2)ds}_{=K^{(1)}_{0,0}=1}\right.$$
$$+\frac{h^2}{2}f_{x_1x_1}(x_1,x_2)\underbrace{\int_{\mathbb{R}^2} s_1^2 K(s_1,s_2)ds}_{=K^{(1)}_{2,0}}$$
$$+\frac{h^2}{2}f_{x_2x_2}(x_1,x_2)\underbrace{\int_{\mathbb{R}^2} s_2^2 K(s_1,s_2)ds}_{=K^{(1)}_{0,2}}$$
$$+\frac{h^4}{24}f_{x_1x_1x_1x_1}(x_1,x_2)\underbrace{\int_{\mathbb{R}^2} s_1^4 K(s_1,s_2)ds}_{=K^{(1)}_{4,0}}$$
$$+\frac{h^4}{4}f_{x_1x_1x_2x_2}(x_1,x_2)\underbrace{\int_{\mathbb{R}^2} s_1^2 s_2^2 K(s_1,s_2)ds}_{=K^{(1)}_{2,2}}$$
$$+\frac{h^4}{24}f_{x_2x_2x_2x_2}(x_1,x_2)\underbrace{\int_{\mathbb{R}^2} s_2^4 K(s_1,s_2)ds}_{=K^{(1)}_{0,4}}$$
$$+O(h^6)\big)\,dx$$

terms which equals to zero are omitted

$$= \frac{K(0,0)}{nh^2} + \frac{n-1}{n}\int_{\mathbb{R}^2}\Big(\big(f(x_1,x_2)\big)^2$$
$$+\frac{h^2}{2}f(x_1,x_2)f_{x_1x_1}(x_1,x_2)K^{(1)}_{2,0} + \frac{h^2}{2}f(x_1,x_2)f_{x_2x_2}(x_1,x_2)K^{(1)}_{0,2}$$
$$+\frac{h^4}{24}f(x_1,x_2)f_{x_1x_1x_1x_1}(x_1,x_2)K^{(1)}_{4,0} + \frac{h^4}{24}f(x_1,x_2)f_{x_2x_2x_2x_2}(x_1,x_2)K^{(1)}_{0,4}$$
$$+\frac{h^4}{4}f(x_1,x_2)f_{x_1x_1x_2x_2}(x_1,x_2)K^{(1)}_{2,2} + f(x_1,x_2)O(h^6)\Big)\,dx$$
$$= \frac{K(0,0)}{nh^2} + \frac{n-1}{n}\Big(I^{(1)}_0$$

$$
\begin{aligned}
&+\frac{K_{2,0}^{(1)}h^2}{2}\int_{I\!R^2} f(x_1,x_2)\left(f_{x_1x_1}(x_1,x_2)+f_{x_2x_2}(x_1,x_2)\right)dx\\
&+\frac{K_{4,0}^{(1)}h^4}{12}\int_{I\!R^2} f(x_1,x_2)\left(f_{x_1x_1x_1x_1}(x_1,x_2)+f_{x_2x_2x_2x_2}(x_1,x_2)\right)dx\\
&+\frac{K_{2,2}^{(1)}h^4}{4}\int_{I\!R^2} f(x_1,x_2)f_{x_1x_1x_2x_2}(x_1,x_2)dx+O(h^6)\Bigg)\\
=\;&\frac{K(0,0)}{nh^2}+\frac{n-1}{n}I_0^{(1)}+\frac{K_{2,0}^{(1)}I_2^{(1)}h^2(n-1)}{2n}\\
&+\frac{K_{4,0}^{(1)}I_4^{(1)}h^4(n-1)}{24n}+\frac{K_{2,2}^{(1)}Jh^4(n-1)}{4n}+O(h^6),
\end{aligned}
$$

and the bias will be

$$
\begin{aligned}
Bias(\hat{I}_h) \;=\;& \frac{K(0,0)}{nh^2}-\frac{I_0^{(1)}}{n}+\frac{K_{2,0}^{(1)}I_2^{(1)}h^2(n-1)}{2n}\\
&+\frac{K_{4,0}^{(1)}I_4^{(1)}h^4(n-1)}{24n}+\frac{K_{2,2}^{(1)}Jh^4(n-1)}{4n}+O(h^6).
\end{aligned}
$$

It follows that the estimator is asymptotically unbiased ($n \longrightarrow \infty$, $h \longrightarrow 0$). The squared bias is

$$
\begin{aligned}
Bias^2(\hat{I}_h) \;=\;& \frac{K(0,0)^2}{n^2h^4}-\frac{2K(0,0)I_0^{(1)}}{n^2h^2}+\frac{K(0,0)K_{2,0}^{(1)}I_2^{(1)}(n-1)}{n^2}\\
&+\frac{K(0,0)K_{4,0}^{(1)}I_4^{(1)}h^2(n-1)}{12n^2}+\frac{K(0,0)K_{2,2}^{(1)}Jh^2(n-1)}{2n^2}\\
&+\frac{I_0^{(1)2}}{n^2}-\frac{I_0^{(1)}K_{2,0}^{(1)}I_2^{(1)}h^2(n-1)}{n^2}+O(h^4)\\
=\;& h^{-4}\frac{K(0,0)^2}{n^2}-h^{-2}\frac{2K(0,0)I_0^{(1)}}{n^2}\\
&+\frac{K(0,0)K_{2,0}^{(1)}I_2^{(1)}(n-1)}{n^2}+\frac{I_0^{(1)2}}{n^2}\\
&+h^2\Bigg(\frac{K(0,0)K_{4,0}^{(1)}I_4^{(1)}(n-1)}{12n^2}-\frac{I_0^{(1)}K_{2,0}^{(1)}I_2^{(1)}(n-1)}{n^2}\\
&+\frac{K(0,0)K_{2,2}^{(1)}Jh^2(n-1)}{2n^2}\Bigg)+O(h^4)
\end{aligned}
$$

and the variance of the estimator is

$$
\begin{aligned}
Var(\hat{I}_h) &= Var\left(\frac{1}{n^2h^2}\left(nK_h(0,0) + 2\sum_{\substack{i,j=1\\ i<j}}^{n} K_h(X_i - X_j)\right)\right) \\
&= \frac{4}{n^4h^4} Var\left(\sum_{\substack{i,j=1\\ i<j}}^{n} K_h(X_i - X_j)\right) \\
&= \frac{4}{n^4h^4}\left(E\left(\left(\sum_{\substack{i,j=1\\ i<j}}^{n} K_h(X_i - X_j)\right)^2\right) - E\left(\sum_{\substack{i,j=1\\ i<j}}^{n} K_h(X_i - X_j)\right)^2\right) \\
&= \frac{4}{n^4h^4} E\left(\sum_{\substack{i,j=1\\ i<j}}^{n}\sum_{\substack{l,m=1\\ l<m}}^{n} K_h(X_i - X_j)K_h(X_l - X_m)\right) \\
&\quad -\frac{4}{n^4h^4}\frac{n^2(n-1)^2}{2} E\left(K_h(X-Y)\right)^2 \\
&= \frac{4}{n^4h^4}\left(\frac{n(n-1)}{2} E\left(K_h(X-Y)^2\right)\right. \\
&\quad +2\frac{n(n-1)(n-2)}{3} E\left(K_h(X-Y)K_h(X-Z)\right) \\
&\quad +\left.\left(\frac{n^2(n-1)^2}{4} - 2\frac{n(n-1)(n-2)}{3} - \frac{n(n-1)}{2}\right)\right. \\
&\quad E\left(K_h(X-Y)K_h(W-Z)\right)) - \frac{2(n-1)^2}{n^2h^4} E\left(K_h(X-Y)\right)^2 \\
&= \frac{2(n-1)}{n^3h^4} \qquad E\left(K_h(X-Y)^2\right) \\
&\quad + \frac{8(n-1)(n-2)}{3n^3h^4} \qquad E\left(K_h(X-Y)K_h(X-Z)\right) \\
&\quad + \frac{(n-1)(n-2)(3n-5)}{3n^3h^4} \qquad E\left(K_h(X-Y)\right)^2 \\
&\quad - \frac{2(n-1)^2}{n^2h^4} \qquad E\left(K_h(X-Y)\right)^2 \\
&= \frac{2(n-1)}{n^3h^4} \qquad E\left(K_h(X-Y)^2\right) \\
&\quad + \frac{8(n-1)(n-2)}{3n^3h^4} \qquad E\left(K_h(X-Y)K_h(X-Z)\right) \\
&\quad - \frac{(n-1)}{n^2h^4}\frac{3n^2+5n+10}{3n} \qquad E\left(K_h(X-Y)\right)^2 .
\end{aligned}
$$

Now we calculate the expectations separately starting with $E(K_h(X-Y)^2)$

$$
\begin{aligned}
&= \int_{\mathbb{R}^2} f(x_1,x_2)\int_{\mathbb{R}^2} K\left(\frac{x_1-y_1}{h},\frac{x_2-y_2}{h}\right)^2 f(y_1,y_2)dydx \\
&= h^2\int_{\mathbb{R}^2} f(x_1,x_2)\int_{\mathbb{R}^2} K(s_1,s_2)^2 f(x_1-hs_1,x_2-hs_2)dsdx \\
&= h^2\int_{\mathbb{R}^2} f(x_1,x_2)\Big(f(x_1,x_2)K^{(2)}_{0,0} \\
&\quad +\frac{h^2}{2}f_{x_1x_1}(x_1,x_2)K^{(2)}_{2,0}+\frac{h^2}{2}f_{x_1x_1}(x_1,z_2)K^{(2)}_{0,2} \\
&\quad +\frac{h^4}{24}f_{x_1x_1x_1x_1}(x_1,x_2)K^{(2)}_{4,0}+\frac{h^4}{24}f_{x_2x_2x_2x_2}(x_1,x_2)K^{(2)}_{0,4} \\
&\quad +\frac{h^4}{4}f_{x_1x_1x_2x_2}(x_1,x_2)K^{(2)}_{2,2}+O(h^6)\Big)\,dx \\
&= h^2K^{(2)}_{0,0}I^{(1)}_0+\frac{h^4}{2}K^{(2)}_{2,0}I^{(1)}_2+\frac{h^6}{24}(K^{(2)}_{4,0}I^{(1)}_4+6K^{(2)}_{2,2}J)+O(h^8) \\
&= h^2K^{(2)}_{0,0}I^{(1)}_0+\frac{h^4}{2}K^{(2)}_{2,0}I^{(1)}_2+\frac{h^6}{24}M_2+O(h^8),
\end{aligned}
$$

then the expectation of $E(K_h(X-Y))$

$$
\begin{aligned}
&= \int_{\mathbb{R}^2} f(x_1,x_2)\int_{\mathbb{R}^2} K\left(\frac{x_1-y_1}{h},\frac{x_2-y_2}{h}\right) f(y_1,y_2)dydx \\
&= h^2\int_{\mathbb{R}^2} f(x_1,x_2)\int_{\mathbb{R}^2} K(s_1,s_2) f(x_1-hs_1,x_2-hs_2)dsdx \\
&= h^2\int_{\mathbb{R}^2} f(x_1,x_2)\,(f(x_1,x_2) \\
&\quad +\frac{h^2}{2}f_{x_1x_1}(x_1,x_2)K^{(1)}_{2,0}+\frac{h^2}{2}f_{x_1x_1}(x_1,x_2)K^{(1)}_{0,2} \\
&\quad +\frac{h^4}{24}f_{x_1x_1x_1x_1}(x_1,x_2)K^{(1)}_{4,0}+\frac{h^4}{24}f_{x_2x_2x_2x_2}(x_1,x_2)K^{(1)}_{0,4} \\
&\quad \frac{h^4}{4}f_{x_1x_1x_2x_2}(x_1,x_2)K^{(1)}_{2,2}++O(h^6)\Big)\,dx \\
&= h^2I^{(1)}_0+\frac{h^4}{2}K^{(1)}_{2,0}I^{(1)}_2+\frac{h^6}{24}(K^{(1)}_{4,0}I^{(1)}_4+6K^{(1)}_{2,2}J)+O(h^8) \\
&= h^2I^{(1)}_0+\frac{h^4}{2}K^{(1)}_{2,0}I^{(1)}_2+\frac{h^6}{24}M_1+O(h^8)
\end{aligned}
$$

and last the expectation of $E(K_h(X-Y)K_h(X-Z))$

$$
= \int_{\mathbb{R}^2}\int_{\mathbb{R}^2} f(x_1,x_2)f(y_1,y_2)K\left(\frac{x_1-y_1}{h},\frac{x_2-y_2}{h}\right)
$$

$$
\begin{aligned}
&\int_{\mathbb{R}^2} K\left(\frac{x_1 - z_1}{h}, \frac{x_2 - z_2}{h}\right) f(z_1, z_2) dz dy dx \\
= & \int_{\mathbb{R}^2} \int_{\mathbb{R}^2} f(x_1, x_2) f(y_1, y_2) K\left(\frac{x_1 - y_1}{h}, \frac{x_2 - y_2}{h}\right) \\
& h^2 \left(f(x_1, x_2) + \frac{h^2}{2} K_{2,0}^{(1)} (f_{x_1 x_1}(x_1, x_2) + f_{x_2 x_2}(x_1, x_2)) \right. \\
& \left. + O(h^4) \right) dy dx \\
= & h^4 \int_{\mathbb{R}^2} f(x_1, x_2) \left(f(x_1, x_2) + \frac{h^2}{2} K_{2,0}^{(1)} (f_{x_1 x_1}(x_1, x_2) + f_{x_2 x_2}(x_1, x_2)) \right. \\
& \left. + O(h^4) \right)^2 dx \\
= & h^4 \int_{\mathbb{R}^2} f(x_1, x_2) \left(f(x_1, x_2)^2 + h^2 K_{2,0}^{(1)} (f_{x_1 x_1}(x_1, x_2) + f_{x_2 x_2}(x_1, x_2)) \right. \\
& \left. + O(h^4) \right) dx \\
= & h^4 I_0^{(2)} + h^6 K_{2,0}^{(1)} I_2^{(2)} + O(h^8).
\end{aligned}
$$

Now we put all expectations together and it follows

$$
\begin{aligned}
Var(\hat{I_h}) = & \frac{2(n-1)}{n^3 h^4} \left(h^2 K_{0,0}^{(2)} I_0^{(1)} + \frac{h^4}{2} K_{2,0}^{(2)} I_2^{(1)} + \frac{h^4}{2} K_{2,0}^{(2)} I_2^{(1)} + \frac{h^6}{24} M_2 + O(h^8) \right) \\
& + \frac{8(n-1)(n-2)}{3n^3 h^4} \left(h^4 I_0^{(2)} + h^6 K_{2,0}^{(1)} I_2^{(2)} + O(h^8) . \right) \\
& - \frac{(n-1)}{n^2 h^4} \left(\frac{3n^2 + 5n + 10}{3n} \right) \\
& \left(h^2 I_0^{(1)} + \frac{h^4}{2} K_{2,0}^{(1)} I_2^{(1)} + \frac{h^6}{24} M_1 + O(h^8) \right)^2 \\
= & \frac{2(n-1)}{n^3 h^4} \left(h^2 K_{0,0}^{(2)} I_0^{(1)} + \frac{h^4}{2} K_{2,0}^{(2)} I_2^{(1)} + \frac{h^4}{2} K_{2,0}^{(2)} I_2^{(1)} + \frac{h^6}{24} M_2 + O(h^8) \right) \\
& + \frac{8(n-1)(n-2)}{3n^3 h^4} \left(h^4 I_0^{(2)} + h^6 K_{2,0}^{(1)} I_2^{(2)} + O(h^8) . \right) \\
& - \frac{(n-1)}{n^2 h^4} \left(\frac{3n^2 + 5n + 10}{3n} \right) \\
& \left(h^4 I_0^{(1)^2} + h^6 K_{2,0}^{(1)} I_2^{(1)} I_0^{(1)} + O(h^8) \right) \\
= & h^{-2} \frac{2(n-1)}{n^3} K_{0,0}^{(2)} I_0^{(1)} \\
& + \frac{2(n-1)}{n^3} K_{2,0}^{(2)} I_2^{(1)} + \frac{8(n-1)(n-2)}{3n^3} I_0^{(2)} \\
& - I_0^{(1)2} \frac{(n-1)(3n^2 + 5n + 10)}{3n^3}
\end{aligned}
$$

$$+h^2\left(\frac{M_2}{24}+\frac{8(n-1)(n-2)}{3n^3}K_{2,0}^{(1)}I_2^{(2)}\right.$$
$$\left.-\frac{(n-1)(3n^2+5n+10)}{3n^3}K_{2,0}^{(1)}I_2^{(1)}I_0^{(1)}\right)+O(h^4)$$

With $n \longrightarrow \infty$, $h \longrightarrow 0$ and $nh \longrightarrow \infty$ the variance converges to ${I_0^{(1)}}^2$. The mean squared error of the Friedman-Tukey index will be

$$\begin{aligned}
MSE(\hat{I}_{FT}) &= Bias^2(\hat{I}_{FT})+Var(\hat{I}_{FT})\\
&= h^{-4}\frac{K(0,0)^2}{n^2}\\
&\quad +h^{-2}\left(-\frac{2K(0,0)I_0^{(1)}}{n^2}+\frac{2(n-1)}{n^3}K_{0,0}^{(2)}I_0^{(1)}\right)\\
&\quad +\frac{K(0,0)K_{2,0}^{(1)}I_2^{(1)}(n-1)}{n^2}+\frac{I_0^{(1)2}}{n^2}\\
&\quad +\frac{2(n-1)}{n^3}K_{2,0}^{(2)}I_2^{(1)}+\frac{8(n-1)(n-2)}{3n^3}I_0^{(2)}\\
&\quad -I_0^{(1)2}\frac{(n-1)(3n^2+5n+10)}{3n^3}\\
&\quad +h^2\left(\frac{K(0,0)K_{4,0}^{(1)}I_4^{(1)}(n-1)}{12n^2}-\frac{I_0^{(1)}K_{2,0}^{(1)}I_2^{(1)}(n-1)}{n^2}\right.\\
&\quad +\frac{K(0,0)K_{2,2}^{(1)}J(n-1)}{2n^2}+\frac{M_2}{24}+\frac{8(n-1)(n-2)}{3n^3}K_{2,0}^{(1)}I_2^{(2)}\\
&\quad \left.-\frac{(n-1)(3n^2+5n+10)}{3n^3}K_{2,0}^{(1)}I_2^{(1)}I_0^{(1)}\right)+O(h^4)\\
&= \frac{C_0}{h^4}+\frac{C_1}{h^2}+C_2+C_3h^2+O(h^4,h^4n^{-1},h^4n^{-2},\ldots)
\end{aligned}$$

with

$$\begin{aligned}
C_0 &= \frac{K(0,0)^2}{n^2}\\
C_1 &= -\frac{2K(0,0)I_0^{(1)}}{n^2}+\frac{2(n-1)}{n^3}K_{0,0}^{(2)}I_0^{(1)}\\
C_2 &= \frac{K(0,0)K_{2,0}^{(1)}I_2^{(1)}(n-1)}{n^2}+\frac{I_0^{(1)2}}{n^2}\\
&\quad +\frac{2(n-1)}{n^3}K_{2,0}^{(2)}I_2^{(1)}+\frac{8(n-1)(n-2)}{3n^3}I_0^{(2)}\\
&\quad -I_0^{(1)2}\frac{(n-1)(3n^2+5n+10)}{3n^3}
\end{aligned}$$

$$
\begin{aligned}
C_3 &= \frac{K(0,0)K_{4,0}^{(1)}I_4^{(1)}(n-1)}{12n^2} - \frac{I_0^{(1)}K_{2,0}^{(1)}I_2^{(1)}(n-1)}{n^2} \\
&\quad + \frac{K(0,0)K_{2,2}^{(1)}J(n-1)}{2n^2} + \frac{M_2}{24} + \frac{8(n-1)(n-2)}{3n^3}K_{2,0}^{(1)}I_2^{(2)} \\
&\quad - \frac{(n-1)(3n^2+5n+10)}{3n^3}K_{2,0}^{(1)}I_2^{(1)}I_0^{(1)}
\end{aligned}
$$

If we plug in the triweight kernel and the bivariate standard gaussian distribution we get for the constants:

$$
\begin{aligned}
K(0,0) &= 1.27324 \\
K_{0,0}^{(2)} &= 0.727565 \\
K_{2,0}^{(1)} &= 0.1 \\
K_{4,0}^{(1)} &= 0.025 \\
K_{4,0}^{(2)} &= 0.0757889 \\
K_{2,2}^{(1)} &= 0.00833333 \\
K_{2,2}^{(2)} &= 0.00252627 \\
I_0^{(1)} &= 0.0503026 \\
I_2^{(1)} &= -0.0795775 \\
I_4^{(1)} &= 0.185235 \\
I_2^{(2)} &= -0.0106299 \\
J &= 0.0308725
\end{aligned}
$$

C

Density Estimation on Hexagonal Bins

As pointed out, e.g. in Klinke (1994), other forms than squares are possible for bins. We can even use tesselations of the plane which consist of different bin types. The most reasonable form of a bin, in connection with bivariate kernel estimation with a circle as support, is the hexagonal bin. In fact the binning algorithm and estimation algorithm become more complicated.

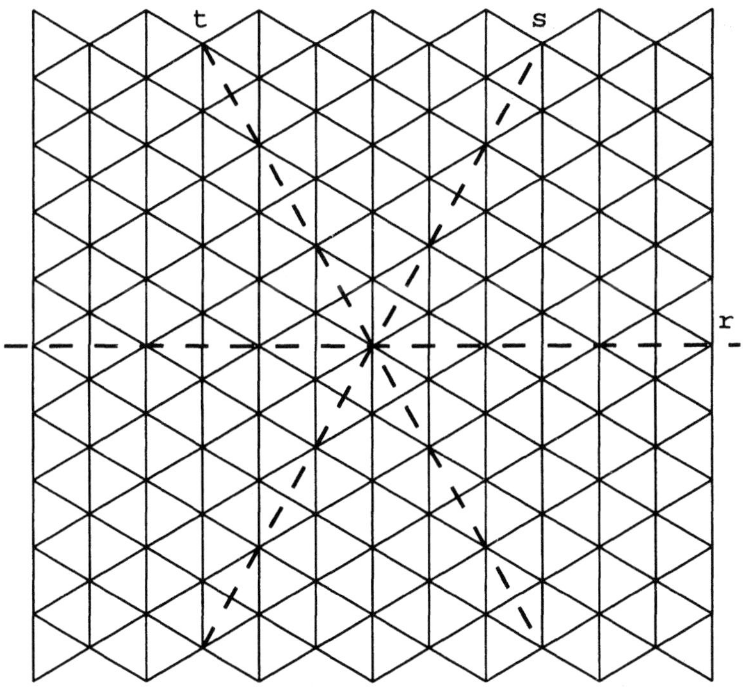

FIGURE C.1. Hexagonal binning with bins built up from triangular bins.

For binning we need to introduce triangular coordinates (r, s, t). So our data $(x_i, y_i)_{i=1,\dots,n}$ will be transformed into

$$\begin{aligned} r_i &= \operatorname{int}\left(\frac{x_i}{\delta}\right) \\ s_i &= \operatorname{int}\left(\frac{\cos(\pi/3)x_i + \sin(\pi/3)y_i}{\delta}\right) \\ t_i &= \operatorname{int}\left(\frac{\cos(2\pi/3)x_i + \sin(2\pi/3)y_i}{\delta}\right) \end{aligned}$$

In the moment the choice of the triangular coordinate axis r parallel to the euclidean axis x seems to be arbitrary. We will see later that this choice is necessary. Let us take the hexagonal bin, which contains the triangular bin $(0,0,0)$, the triangular bin in s-direction from the cross-over of the 3 axes. We see that the bins $(0,0,0), (0,0,-1), (0,-1,-1), (-1,-1,-1), (-1,-1,0)$ and $(-1,0,0)$ form our hexagonal bin $\{0,0\}$. A shift in direction of r results in an addition of $(2,1,-1)$ to these bins to get the triangular bins of the hexagonal bin $\{1,0\}$. To get the hexagonal bin $\{0,1\}$ we have to add $(1,2,1)$ to the triangular bins of $\{0,0\}$. Now we can easily calculate the triangular bins which belong to the hexagonal bin $\{l,m\}$ by adding $(2l+m, l+m, -l+m)$ to the triangular bins of the hexagon $\{0,0\}$. As a consequence we can derive that

$$r_i - s_i + t_i = \begin{cases} 0 \\ -1 \end{cases}$$

But we are interested in the other way, that means which datapoint belongs to which hexagonal bin. The equations we have to solve are:

$$(r_i, s_i, t_i) - l(2,1,-1) - m(1,2,1) = \begin{cases} (0,0,0) \\ (0,0,-1) \\ (0,-1,-1) \\ (-1,-1,-1) \\ (-1,-1,0) \\ (-1,0,0) \end{cases}$$

With the relationship for $r_i - s_i + t_i$ and the knowledge that l and m are integer the following algorithm gives the solution:

```
if (ti-si+ri)                       // ti-si+ri = -1
{ p = 2+2*ri-si;                    // check (-1,0,0)
  q = -1-ri+2*si;
  if ((p%3) || (q%3))
  { p = 1+2*ri-si;                  // check (-1,-1,0)
```

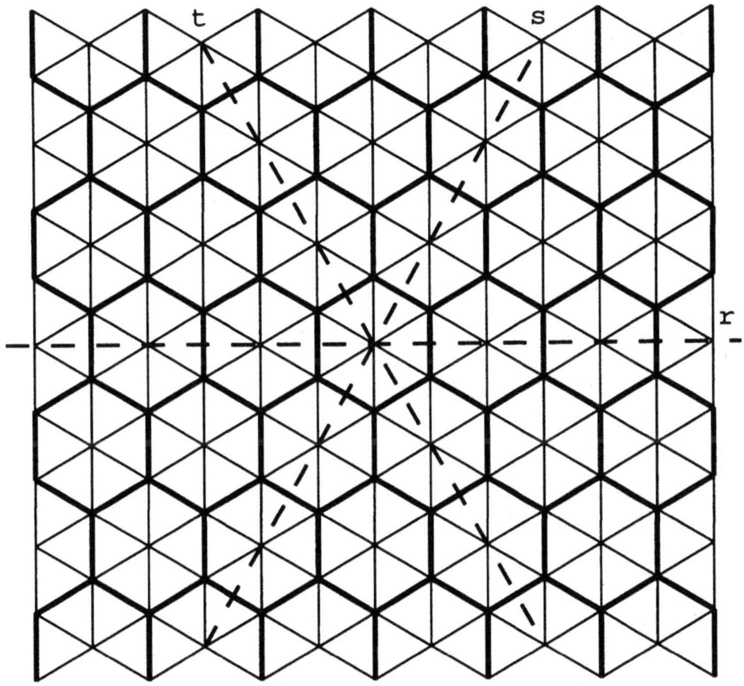

FIGURE C.2. Hexagonal binning with bins built up from triangular bins.

```
    q = 1-ri+2*si;
    if ((p%3) || (q%3))
    { p = 2*ri-si;              // check (0,0,-1)
      q = -ri+2*si;
    }
  }
}
else                            // ti-si+ri = 0
{ p = 2*ri-si;                  // check (0,0,0)
  q = -ri+2*si;
  if ((p%3) || (q%3))
  { p = 1+2*ri-si;              // check (-1,-1,0)
    q = 1-ri+2*si;
    if ((p%3) || (q%3))
    { p = -1+2*ri-si;           // check (0,-1,-1)
      q = 2-ri+2*si;
    }
  }
```

```
}
li = p/3;
mi = q/3;
```

We have now expressed each datapoint (x_i, y_i) in coordinates of the hexagonal bins $\{l_i, m_i\}$. If we want to calculate e.g. the Friedman-Tukey index on hexagonal bins, we have mainly to calculate kernel of distances. As in the case of squared bins we store the values of the kernel $K\{l2\delta, m2\delta\}$. But l and m will not run from 0 up to $ceil(h/\delta)$. Similar to squared bins, where we restrict ourselves to the upper quarter $(l \geq 0, m \geq 0)$, we can here restrict ourselves to the hexagons enclosed by the r and s axis. So we calculate $K\{l2\delta, m2\delta\}$ with $b = ceil(h/(2\delta\cos(\pi/6)))$ $0 \leq l, m \leq b$ with the constraint that $l + m \leq b$:

```
int i, j, b, k0;
double *k;
b  = ceil (bandwidth / (2.0 * binwidth * cos(pi/6.0)));
f  = 2.0 * binwidth * cos(pi/6.0) / bandwidth
k  = malloc (b*b*sizeof(double));
for (l=0; i<b; i++)
{ for (m=0; j<b; j++)
    if (l+m<b)
      *(k+l*b+m) = kern (f * (l + m * cos(pi/6)),
                         f * l * sin(pi/6));
    else
      *(k+l*b+m) = 0.0
}
```

The subroutine *kern* has to give back the values of $K(x, y)$. Some simplifications can be made for the case that the kernel is known. One of the easiest methods to accelerate the speed of the kernel indices in the case of squared bins is the use of the symmetry of the kernel and to sort the data after the first component. An efficient program would be

```
double index;
int i, j, start;
index = 0.0;
start = 0;
for (i=0; i<n; i++)
{ for (j=start; j<i; j++)
  { dx = *(l+i) - *(l+j);  // always greater equal zero
    if (dx<b)
    { dy = abs(*(m+i) - *(m+j));
      if (dy<b) index += *(k+dx*b+dy);
```

```
      }
      else
        start++;
    }
  }
  index = (*k + 2.0 * index) / (n*bandwidth*bandwidth)
```

In general we can use the same technique for the hexagonal bins, but the mapping from the other quarters to the left upper quarter in the case of squared bins is much easier. Since the data are sorted through the first variable x_i we know that for every x_i the x_j $j = 1, ..., i-1$ has to be left of it. The symmetry of the kernel ensures that we can use the distance $x_i - x_j$ instead of $x_j - x_i$, so we only need to make mappings for l, m with $l \geq 0, m$ arbitrary and $l \leq 0, m \geq -l$. This results in four parts, where one is exactly the area for which we calculated the kernel. The other areas have to be mapped by

$$\begin{array}{lll} l \leq 0, & -l \leq m \leq 0 & \rightarrow \{-l, l+m\} \\ l \geq 0, & m \geq 0 & \rightarrow \{l, m\} \\ l \geq 0, & -l \leq m \leq 0 & \rightarrow \{l+m, -m\} \\ l \geq 0, & m \leq -l & \rightarrow \{l, -m-l\} \end{array}$$

The following program realizes it:

```
double index;
int i, j, mx,my;
index = 0.0;
for (i=0; i<n; i++)
{ for (j=1; j<i; j++)
  { dx = *(l+i) - *(l+j);
    dy = *(m+i) - *(m+j);
    if (dx<0)
    { mx = -dx; my = dx+dy; }
    else if (dy>=0)
    { mx = dx; my = dy}
    else if (dy>=-dx}
    { mx = dx+dy; my = -dy; }
    else
    { mx = dx; my = -dx-dy; }
    if ((mx<b) && (my<b)) index += *(k+mx*b+my);
  }
}
index = (n * *k + 2.0 * index) / (n*bandwidth*bandwidth)
```

Although this is short, it will need more time than the program for squared bins:

1. We can not use a window running over the data as in the case of squared bins
2. We have to do the mapping by a `if` sequence being of intensive computing time, but in the case of squared bins we could use the `abs` function which is less time intensive

The conclusion is that it is possible to bin and to estimate on hexagonal bins, which obviously leads to a more accurate estimate, but in terms of computational time it is worse.

D

Programs

D.1 XploRe Programs

Most of the pictures were done with the help of XploRe programs which are available via WWW:

`http://www.wiwi.hu-berlin.de/~sigbert/thprog.html`

- **ANDREW/** — Andrews curves
  - **ANDREW/ANDREW.XPL**
- **BOXPLOT/** — Boxplots
  - **BOXPLOT/BOXCOMP.XPL** — Comparing boxplots
  - **BOXPLOT/BOXSUB.XPL** — Subgroup analysis
- **CLUSTER/** — Cluster analysis
  - **CLUSTER/CLDIST.MNU** — Menu for distances
  - **CLUSTER/CLUSREST.XPL** — Routines for clustering
  - **CLUSTER/CLUSTER.MNU** — Main menu
  - **CLUSTER/CLUSTER.XPL** — Interface routine
  - **CLUSTER/CLWEIGHT.MNU** — Menu for weights
- **DATA/** — Data
  - **DATA/BANK2.DAT** — Swiss banknote dataset
  - **DATA/FLAT1094.DAT** — Berlin flat data (offers 10/94)
  - **DATA/FLAT8994.EXE**
    Self extracting archive with the Berlin flat data (all offers)
  - **DATA/OBOST.DAT** — Boston housing data
  - **DATA/XWAVELET.DAT** — Constants for generating wavelets (X-format)
- **DRAFTMAN/** — Scatterplotmatrix (Splom)
  - **DRAFTMAN/DRAFT4.XPL** — Splom for 4D-data
  - **DRAFTMAN/DRAFT5.XPL** — Splom for 5D-data
  - **DRAFTMAN/DRAFT6.XPL** — Splom for 6D-data
  - **DRAFTMAN/DRAFT7.XPL** — Splom for 7D-data
  - **DRAFTMAN/DRAFTINF.XPL** — Splom > 7D-data

- DRAFTMAN/DRAFTMAN.XPL — Interface routine

- EPP/ — Exploratory projection pursuit
  - EPP/PPE.XPL — Routine for EPP
  - EPP/PPEXPL.XPL — Interface routine
  - EPP/PPGRAY.MNU — Menu for colorselection
  - EPP/PPINTER.XPL — Interpretation of EPP
  - EPP/PPMAIN.MNU — Main menu for EPP
  - EPP/PPNORM.MNU — Menu for centering/sphering
- FACE/ — Flury faces
  - FACE/ASSOCIAT.XPL — Association between face part and variables
  - FACE/ASYMFACE.MNU — Selection for asymmetric faces
  - FACE/DOFACE.XPL — Drawing faces
  - FACE/SYMFACE.MNU — Selection for symmetric faces
- GENERATE/ — Random generator
  - GENERATE/NORMGEN.XPL
- HISTOGRM/ — Histograms
  - HISTOGRM/ASH.XPL — Average shifted histogram with uniform kernel
  - HISTOGRM/ASHK.XPL — ASH with arbitratry kernel
  - HISTOGRM/HISTCOMP.XPL — Comparing histograms
  - HISTOGRM/HISTOGRM.XPL — Generating histograms
- KERDENS/ — Kernel Density
  - KERDENS/DENEST.XPL — Univariate kernel density (binned)
  - KERDENS/DENESTP.XPL — Multivariate kernel density (binned)
  - KERDENS/DENS3PLT.XPL — Plotting of marginal densities
  - KERDENS/SKERDENS.XPL — Multivariate kernel density (unbinned)
  - KERDENS/SYMWEIGH.XPL — Rescaling of kernels
- KERNEL/ — Kernel
  - KERNEL/GAU.XPL — Gaussian kernel
  - KERNEL/QUA.XPL — Quartic product kernel
  - KERNEL/REPA.XPL — Radial epanechnikov kernel
  - KERNEL/RQUA.XPL — Radial quartic kernel
  - KERNEL/RTRI.XPL — Radial triweight kernel
  - KERNEL/RTRIAN.XPL — Radial triangle kernel
  - KERNEL/RUNI.XPL — Radial uniform kernel
  - KERNEL/TRI.XPL — Triweight kernel

- KERREG/ — Kernel regression
  - KERREG/REGESTP.XPL — Multivariate kernel regression (binned)
  - KERREG/SKERREG.XPL — Multivariate kernel regression (unbinned)
  - KERREG/SYMWEIGH.XPL — Rescaling of kernels
- MDS/ — Multidimensional Scaling
  - MDS/DOMDS.XPL — Plotting of metric scaling
  - MDS/MDS.XPL — Metric scaling
  - MDS/NMDS.XPL — Nonmetric scaling
- PCA/ — Principal component analysis
  - PCA/HDPCA.XPL
- PCP/ — Parallel coordinate plot
  - PCP/PCPLOT.XPL
- PPR/ — Projection pursuit regression
  - PPR/PPR.XPL — Interface routine
  - PPR/PPRINTER.MNU — Menu for interpreting
  - PPR/PPRINTER.XPL — Routine for interpreting
  - PPR/PPRMAIN.MNU — Main menu for PPR
  - PPR/PPRREST.XPL — Routines for PPR
- PROG/ — Several programs
  - PROG/DESKSTAT.TXT — Text for desc. statistics
  - PROG/DESKSTAT.XPL — Computing descriptive statistics
  - PROG/FIVENUM.XPL — Five number summary
  - PROG/JITTER.XPL — Jittering of data
  - PROG/LINREG.XPL — Linear regression
  - PROG/LOCREG.XPL — Local regression
  - PROG/SUNFLOW.XPL — Sunflower plot
  - PROG/VARCOMP.XPL — Comparing variables
- SAS/ — SAS program
  - SAS/LINREG.SAS
- SCATTER/ — Scatterplot
  - SCATTER/S2MAIN.MNU — Main menu
  - SCATTER/S2REGR.MNU — Menu for regression
  - SCATTER/S2REST.XPL — Routines for scatterplot
  - SCATTER/S2TRF.MNU — Menu for variable transformation

- `SCATTER/S2VIEW.MNU` — Menu for views
- `SCATTER/SCATTER2.XPL` — Interface routine

- `TEACHWAR/` — Teachware
- `TOUR/` — Grand tour
  - `TOUR/TOUR.XPL` — Interface routine
  - `TOUR/TOURADD.XPL` — Routine for grand tour
- `WAVELET/` — Wavelet estimation
  - `WAVELET/THRESDEN.MNU` — Menu for threshold in density est.
  - `WAVELET/THRESREG.MNU` — Menu for threshold in regression est.
  - `WAVELET/WAVEDENS.XPL` — Wavelet density estimation
  - `WAVELET/WAVELET.MNU` — Menu for wavelet regression (local)
  - `WAVELET/WAVELET1.XPL` — Interface routine for wavelet regression (local)
  - `WAVELET/WAVELET2.XPL` — Routine for wavelet regression (local)
  - `WAVELET/WAVEREG.XPL` — Wavelet regession estimation
  - `WAVELET/WAVEREGM.MNU` — Main menu for wavelet regession est.
  - `WAVELET/WGEN.XPL` — Generation of father and mother wavelet

D.2 Mathematica Program

D.2.1 Calculation of $p_i(r)$, a_i, b_i and c_i

The following `Mathematica` program calculates the quantities $p_i(r)$, a_i, b_i and c_i. The polynomials p_i are represented as a list of coefficients. The list element `l[i]` is the coefficient before r^{i+1}. The following translations has to be done:

`max-1`	highest p_i that can be calculated
`x[i]`	coefficient list of r^i
`h[i]`	$p_i(r)$
`s[f,g]`	scalarproduct $\int_0^\infty f(r)g(r)w(r)dr$
`intres[i]`	$\int_0^\infty r^i w(r)dr$
`a[i]`	a_i
`b[i]`	b_i
`c[i]`	c_i

This algorithm can be used to calculate a orthogonal function system based on polynomials and specialized scalarproducts. We have only to replace the result of `intres`.

```
max = 15
x[i_] := x[i] = Table[If[k==i, 1, 0], {k, 0, max-1}]
h[0] = x[0]
g[i_] := g[i] = Factor[x[i] - Sum[Simplify[s[x[i],
                h[k]] h[k]], {k, 0, i-1}]]
s[f_, g_] := s[m [f, g]]
s[f_] := f . intvec
intvec = Table[intres[i], {i, 0, max-1}]
intres[0] = 1
intres[1] = p
intres[i_] := intres[i-1] (p+2 i-2)
m[f_,g_] := adjust[CoefficientList[Expand[poly[f] poly[g]], x]]
adjust[f_] := If [Length[f]>=max, Take[f, {1, max}],
              gl = f; While[Length[gl]<max, gl = Append[gl, 0]]; gl]
poly[f_] := f . px
px = Table[x^i, {i, 0, max-1}]
const[i_] := const[i] = norm[g[i]]
norm[f_] := PowerExpand[Sqrt[Factor[s[f,f]]]]
h[i_] := g[i]/norm[g[i]]
a[i_] := Cancel[1/s[h[i], m[x[1], h[i-1]]]]
b[i_] := -Factor[Cancel[a[i] s[h[i-1], m[x[1], h[i-1]]]]]
c[i_] := -Factor[Cancel[a[i] s[h[i-2], m[x[1], h[i-1]]]]]
```

E

Tables

Number	District
1	Frohnau
2	Schildow
3	Buch
4	Heiligensee
5	Berlin 28, Borsigwalde, Hermsdorf, Reinickendorf, Waidmannslust, Wittenau
6.1	Märkisches Viertel, Lübars
6.2	Blankenfelde, Niederschönhausen, Pankow, Rosenthal, Wilhelmsruh
7	Blankenburg, Buchholz, Karow, Wartenberg, Weissensee
8	Spandau
9	Berlin 20, Hakenfelde, Konradshöhe, Spandau
10	Schäfersee, Tegel
11	Berlin 65
11.1	Wedding
12	Hohenschönhausen, Malchow
13	Ahrensfelde, Bernau
14	Falkensee, Staaken
15	Corbusier-Haus, Haselhorst, Olympiastadion, Pichelsdorf, Pichelssee, Scharfe Lanke, Stößensee
16	Berlin 21, Charlottenburg, City, KaDeWe, Kaiser-Friedrich-Straße, Kurfürstendamm, Lietzensee, Moabit, Momsenstrasse, Olivaer Platz, Siemensstadt, Uhlandstrasse, Westend
17.1	Tiergarten
17.2	Mitte
18	Friedrichsfelde, Friedrichshain, Lichtenberg, Prenzlauer Berg
19	Biesdorf, Hellersdorf, Kaulsdorf, Mahlsdorf, Marzahn
20	Dahlwitz-Hoppegarten, Strausberg
21	Gatow
22	Berlin 33, Grunewald, Roseneck
23	Bayerisches Viertel, Friedenau, Halensee, Hohenzollerdamm, Prinzregentenstrasse, Schmargendorf, Viktoria-Luise-Platz, Wexstrasse, Wilmersdorf
24	Berlin 30, Berlin 62, Hasenheide, Kreuzberg, Rixdorf, Schöneberg
25.1	Neukölln
25.2	Baumschulenweg, Karlshorst
26	Treptow
27	Hellersdorf
28	Groß-Glienicke, Havelland-Kladow, Kladow, Schwanenwerder
29	Nikolassee, Schlachtensee, Zehlendorf
30	Berlin 41, Dahlem, Lichterfelde, Schloßstrasse, Steglitz, Viktoriapark, Viktoriaplatz
31	Bayernring, Lankwitz, Marendorf, Tempelhof
32.1	Britz, Bundesgartenschau
32.2	Johannisthal, Niederschöneweide, Oberschöneweide, Treptow
33	Adlershof, Dammvorstadt, Spindlersfelde
34	Friedrichshagen, Hessenwinkel, Köpenick, Müggelsee, Rahnsdorf, Wilhelmshagen
35	Erkner
36	Wahnsee
37	Kleinmachnow
38	Teltow
39	Lichtenrade, Mariendorf
40.1	Buckow, Gropius-Stadt, Rudow
40.2	Altglienicke
41	Falkenhorst, Grünau, Karolinenhof
42	Müggelheim
43	Gosen
44	Schmöckwitz

TABLE E.1. Coding of the location of the Berlin housing data

Time District	1989 7	1989 10	1990 1	1990 4	1990 7	1990 10	1991 1	1991 4	1991 7	1991 10
1	8	9	8			2	3	11	12	12
4	2	1	1	2	3	3		1		2
5	28	20	12	23	30	37	30	42	29	39
6.1	1		1	1				1	1	
8										
9	22	22	21	1	38	45	32	62	36	37
10	11	30	14	4	1	1	1	3	2	6
11.1	11		3	23	18	27	23	16	20	15
14	2	2		1	1			1	3	1
15	31	42	9	36	12	7	9	11	7	15
16	135	157	100	85	77	91	70	118	76	98
17.1	12			3	12	22	11	28	20	5
21							1	1	1	2
22	32	25	15	22	13	4	6	17	10	16
23	65	86	69	59	61	58	44	74	58	51
24	50	70	28	37	28	32	20	51	31	34
25.1	9	2	2	15	8	9	18	25	10	24
28					2	1				1
29	14	18	12	11	21	16	26	33	6	6
30	86	90	40	47	59	57	65	75	66	46
31	20	29	34	34	31	43	44	59	40	39
32.1	10	4	9	6	5	5	4	20	11	13
36	12	6	3	7	5			2		1
37										
38										
39	16	16	20	9	20	19	20	24	28	37
40.1	13	21	12	24	32	39	18	19	25	33
2										
3							1		1	
6.2										
7										2
11.2									1	
12										
13					1					
17.2							1	3		
18										1
19										
20										
25.2										
26										
27										
32.2										
33										
34										
35										
40.2										1
41										
42										
43										
44										

TABLE E.2. Absolute frequencies separated by time and location. The table is splitted up vertically in west/east part and then from west to east.

Time	1992				1993				1994			
District	1	4	7	10	1	4	7	10	1	4	7	10
1	12	14	17	26	2	11	16	21	10	9	15	13
4			4	2			1	1		5	2	3
5	30	47	42	54	32	49	53	52	30	53	77	74
6.1				1			1	3				1
8												
9	22	64	53	70	32	63	60	54	47	58	127	113
10	1	4	6	6	4	4	9	11	3	5	9	10
11.1	10	33	34	37	16	31	27	33	22	32	53	55
14	1	2	1	1	2	2	1	1		2	5	1
15	12	15	19	4		2	3	9	3	19	22	40
16	60	120	149	181	82	163	138	161	65	86	182	218
17.1	33	38	26	39	22	24	37	35	17	14	47	42
21	3	4		1		3	1	2	1			1
22	6	8	12	20	3	18	17	12	12	13	20	26
23	58	82	104	139	61	118	85	88	42	78	147	176
24	28	50	68	90	38	67	81	83	42	37	94	122
25.1	12	29	24	38	21	36	36	21	14	17	43	46
28		2	1			2		2	1	1	9	5
29	7	12	28	32	8	23	34	25	15	22	31	39
30	37	69	82	108	39	89	91	79	31	62	126	140
31	41	48	48	90	39	74	74	66	27	45	116	92
32.1	9	14	19	16	10	18	15	16	11	14	21	23
36	1	5	8	6	2	4	4	4	3	2	18	12
37				1								
38												
39	11	23	13	26	8	28	32	33	13	12	49	32
40.1	22	14	12	29	12	27	37	20	12	17	31	31
2												
3								1			1	
6.2			1	7		3	4	5		10	18	21
7	2			1		1			1	6	12	9
11.2												
12							1		1	1		
13								3				
17.2		1		2					1		2	2
18								2	1	12	13	7
19		1	2	1		4		2			4	5
20						1						
25.2							2	5	2	1	6	3
26												
27												
32.2							1					2
33							1	1				
34			1			2			2	2	4	3
35												
40.2		1	2	1				4	1	2	2	
41						1			6		2	
42												
43												
44		3										

TABLE E.3. Absolute frequencies separated by time and location. The table is splitted up vertically in west/east part and then from west to east.

Value of T	n	T	FA	FL	FR	FB	FM	FP	FI	FE
8907	590	1	198	22	13	2	2	279	279	1
8910	650	1	197	19	14	2	2	299	297	1
9001	413	1	168	20	12	2	2	236	235	1
9004	450	1	169	21	14	2	2	242	242	1
9007	478	1	158	22	12	2	2	230	229	2
9010	518	1	162	20	12	2	2	249	247	1
9101	447	1	149	21	12	2	2	222	221	2
9104	697	1	179	24	14	2	2	278	278	2
9107	494	1	157	23	11	2	2	227	226	2
9110	537	1	171	26	12	2	2	269	268	2
9201	420	1	141	23	12	2	2	212	211	2
9204	702	1	197	26	13	2	2	292	290	2
9207	775	1	241	25	15	2	2	328	328	2
9210	1029	1	270	29	14	2	2	388	386	2
9301	433	1	126	19	12	2	2	220	219	1
9304	870	1	140	28	14	2	2	336	334	2
9307	862	1	230	28	14	2	2	354	350	2
9310	855	1	245	32	13	2	2	348	344	2
9401	436	1	170	29	12	2	2	238	238	2
9404	637	1	218	29	13	2	2	325	324	2
9407	1308	1	329	32	13	2	2	464	457	2
9410	1367	1	366	32	15	2	2	487	476	2

TABLE E.4. Stratification after time of the Berlin flat data.

Value of T	n	DI	DF	DU	DR	DW	DS	DN	DB	TI
8907	590	10	10	9	10	10	10	11	15	1
8910	650	8	8	8	8	8	8	9	13	1
9001	413	9	9	1	9	9	9	10	14	1
9004	450	10	10	1	10	10	10	11	14	1
9007	478	11	11	2	11	11	11	11	15	1
9010	518	10	10	1	10	10	10	10	14	1
9101	447	12	12	2	12	12	12	11	16	1
9104	697	11	11	2	11	11	11	12	16	1
9107	494	12	12	2	12	12	12	11	16	1
9110	537	13	13	2	13	13	13	13	16	1
9201	420	11	11	2	11	11	11	14	16	1
9204	702	14	14	2	14	14	14	14	17	1
9207	775	14	14	2	14	14	14	15	16	1
9210	1029	15	15	2	15	15	15	15	17	1
9301	433	10	10	1	10	10	10	13	15	1
9304	870	15	15	2	15	15	15	15	19	1
9307	862	13	13	2	13	13	13	15	18	1
9310	855	16	16	2	16	16	16	16	18	1
9401	436	16	16	16	16	16	16	15	19	1
9404	637	16	16	16	16	16	16	16	17	1
9407	1308	17	17	17	17	17	17	16	18	1
9410	1367	17	17	17	17	17	17	16	17	1

TABLE E.5. Stratification after time of the Berlin flat data.

Value of DR	n	T	FA	FL	FR	FB	FM	FP	FI	FE
5.98	28	7	24	3	5	2	2	27	27	1
6.54	2130	22	340	2	16	2	2	533	1434	1
7.02	544	22	162	2	13	2	2	221	454	1
8.02	1304	22	274	5	11	2	2	464	944	1
8.90	1483	22	253	5	12	2	2	433	1000	1
8.96	1181	22	260	1	14	2	2	383	854	1
10.29	12	7	12	1	4	2	2	12	12	1
10.50	7	5	7	2	5	2	1	7	7	1
10.57	36	6	29	1	5	2	2	32	36	1
11.39	2718	22	404	3	14	2	2	636	1594	1
11.56	1	1	1	1	1	1	1	1	1	1
13.04	70	9	57	2	6	2	2	58	68	1
14.73	38	10	26	4	5	2	2	34	36	1
14.87	1721	22	301	4	10	2	2	540	1174	1
15.23	38	11	31	2	4	2	2	34	38	1
17.26	19	7	19	1	3	2	2	18	19	1
18.63	2612	22	403	1	16	2	2	614	1632	1
19.13	539	21	136	1	8	2	2	247	427	1
23.71	487	20	140	1	12	2	2	245	407	1

TABLE E.6. Stratification after recreation area of the Berlin flat data.

Value of DR	n	DI	DF	DU	DR	DW	DS	DN	DB	TI
5.98	28	2	2	3	1	1	1	1	3	7
6.54	2130	4	3	7	1	1	1	4	2	22
7.02	544	4	4	7	1	1	2	4	2	22
8.02	1304	3	4	8	1	1	1	7	5	22
8.90	1483	4	4	7	1	1	2	8	5	22
8.96	1181	4	4	6	1	1	3	2	1	22
10.29	12	3	3	3	1	1	2	2	1	7
10.50	7	2	2	3	1	1	2	2	2	5
10.57	36	2	2	3	1	1	1	2	1	6
11.39	2718	4	4	7	1	1	1	4	3	22
11.56	1	1	1	1	1	1	1	1	1	1
13.04	70	3	3	4	1	1	1	2	2	9
14.73	38	3	3	4	1	1	2	3	4	10
14.87	1721	4	4	8	1	1	2	6	3	22
15.23	38	3	3	4	1	1	1	3	2	11
17.26	19	2	2	3	1	1	1	1	1	7
18.63	2612	4	4	6	1	1	1	2	1	22
19.13	539	4	3	6	1	1	2	2	1	21
23.71	487	4	4	6	1	1	1	2	1	20

TABLE E.7. Stratification after recreation area of the Berlin flat data.

```
*************************** PROXIMITIES ***************************
>Warning # 14783
>Due to missing data, some cases have been excluded from computations.

 Data Information
        118 unweighted cases accepted.
         51 cases rejected because of missing value.
 Squared Euclidean measure used.
-----------------------------------------------------------------
******************* HIERARCHICAL CLUSTER ANALYSIS *******************
```

Agglomeration Schedule using Ward Method

	Clusters	Combined		Stage Cluster	1st Appears	Next
Stage	Cluster 1	Cluster 2	Coeff.	Cluster 1	Cluster 2	Stage
1	82	107	2,0	0	0	38
2	80	104	7,5	0	0	7
3	32	108	14,0	0	0	15
4	65	106	21,5	0	0	34
5	67	110	33,5	0	0	17
6	22	109	46,0	0	0	20
7	80	111	60,5	2	0	70
8	30	64	78,5	0	0	66
9	76	116	97,0	0	0	41
10	37	98	115,5	0	0	42
11	89	101	136,5	0	0	49
12	83	93	157,5	0	0	104
13	18	88	178,5	0	0	61
14	29	42	201,5	0	0	86
15	2	32	225,0	0	3	33
16	41	56	250,5	0	0	67

5 pages further

```
Case 20     1   -+            |                                     |
Ca1920     66   -+-+          +-------------------------------------+
Ca2180     74   -+ |          |
Case 859   31   -+ |          |
Ca1739     59   -+ +----------+
Case 620   26   -+ |
Ca1347     48   -+ |
Case 146    7   -+-+
Ca2727     86   -+
Ca1819     62   -+
Case 171    8   -+
Case 336   17   -+
Ca2413     79   -+
```

FIGURE E.1. Cluster analysis of the women labour. One variable contains some missings. Since we have produced a lot of output the information that ~ 30% of the cases are not used can be easily overviewed.

Value of TI	n	T	FA	FL	FR	FB	FM	FP	FI	FE
5.45	436	1	170	29	12	2	2	238	238	2
5.94	855	1	245	32	13	2	2	348	344	2
6.20	637	1	218	29	13	2	2	325	324	2
6.55	862	1	230	28	14	2	2	354	350	2
6.60	870	1	140	28	14	2	2	336	334	2
6.80	590	1	198	22	13	2	2	279	279	1
6.90	1308	1	329	32	13	2	2	464	457	2
7.11	433	1	126	19	12	2	2	220	219	1
7.18	650	1	197	19	14	2	2	299	297	1
7.55	1367	1	366	32	15	2	2	487	476	2
7.60	1029	1	270	29	14	2	2	388	386	2
7.65	413	1	168	20	12	2	2	236	235	1
8.24	775	1	241	25	15	2	2	328	328	2
8.28	702	1	197	26	13	2	2	292	290	2
8.35	420	1	141	23	12	2	2	212	211	2
8.50	697	1	179	24	14	2	2	278	278	2
8.60	537	1	171	26	12	2	2	269	268	2
8.72	494	1	157	23	11	2	2	227	226	2
8.75	450	1	169	21	14	2	2	242	242	1
8.80	478	1	158	22	12	2	2	230	229	2
9.10	447	1	149	21	12	2	2	222	221	2
9.17	518	1	162	20	12	2	2	249	247	1

TABLE E.8. Stratification after interest rate of the German Bundesbank of the Berlin flat data.

Value of TI	n	DI	DF	DU	DR	DW	DS	DN	DB	TI
5.45	436	16	16	16	16	16	16	15	19	1
5.94	855	16	16	2	16	16	16	16	18	1
6.20	637	16	16	16	16	16	16	16	17	1
6.55	862	13	13	2	13	13	13	15	18	1
6.60	870	15	15	2	15	15	15	15	19	1
6.80	590	10	10	9	10	10	10	11	15	1
6.90	1308	17	17	17	17	17	17	16	18	1
7.11	433	10	10	1	10	10	10	13	15	1
7.18	650	8	8	8	8	8	8	9	13	1
7.55	1367	17	17	17	17	17	17	16	17	1
7.60	1029	15	15	2	15	15	15	15	17	1
7.65	413	9	9	1	9	9	9	10	14	1
8.24	775	14	14	2	14	14	14	15	16	1
8.28	702	14	14	2	14	14	14	14	17	1
8.35	420	11	11	2	11	11	11	14	16	1
8.50	697	11	11	2	11	11	11	12	16	1
8.60	537	13	13	2	13	13	13	13	16	1
8.72	494	12	12	2	12	12	12	11	16	1
8.75	450	10	10	1	10	10	10	11	14	1
8.80	478	11	11	2	11	11	11	11	15	1
9.10	447	12	12	2	12	12	12	11	16	1
9.17	518	10	10	1	10	10	10	10	14	1

TABLE E.9. Stratification after interest rate of the German Bundesbank of the Berlin flat data.

Time	1989		1990				1991				1992				1993				1994			
District	7	10	1	4	7	10	1	4	7	10	1	4	7	10	1	4	7	10	1	4	7	10
1	48	48	48			48	48	48	48	48	25	25	25	25	25	25	25	25	25	25	25	25
3							39	39										25		25		
4	48	48	48	48	48	48		48		48	32		32	32			32	32		32	32	32
5	51	51	51	51	51	51	51	51	51	51	33	33	33	33	33	33	33	33	33	33	33	33
6.1	51		51	51				51	51					40			40	40				40
6.2													40	40		40	40	40		40	40	40
7										39	40			40		40			40	40	40	40
8																						
9	51	51	51	51	51	51	51	51	51	51	40	40	40	40	40	40	40	40	40	40	40	40
10	60	60	60	60	60	60	60	60	60	60	40	40	40	40	40	40	40	40	40	40	40	40
11.1	75		75	75	75	75	75	75	75	75	50	50	50	50	50	50	50	50	50	50	50	50
11.2									75													
12																	60		60	60		
13					60													52				
14	39	39		39	39			39	39	39	37	37		37	37	37	37	37		37	37	37
15	57	57	57	57	57	57	57	57	57	57	37	37	37	37		37	37	37	37	37	37	37
16	60	60	60	60	60	60	60	60	60	60	57	57	57	57	57	57	57	57	57	57	57	57
17.1	69			69	69	69	69	69	69	69	47	47	47	47	47	47	47	47	47	47	47	47
17.2							69	69				47		47					47		47	47
18										101								52	52	52	52	52
19												60	60	60		60		60			60	60
20																49						
21							53	53	53	53	29	29		29		29	29	29	29			29
22	39	39	39	39	39	39	39	39	39	39	29	29	29	29	29	29	29	29	29	29	29	29
23	63	63	63	63	63	63	63	63	63	63	44	44	44	44	44	44	44	44	44	44	44	44
24	68	68	68	68	68	68	68	68	68	68	47	47	47	47	47	47	47	47	47	47	47	47
25.1	78	78	78	78	78	78	78	78	78	78	50	50	50	50	50	50	50	50	50	50	50	50
28					39	39				39				38		38	38		38	38	38	38
29	57	57	57	57	57	57	57	57	57	57	40	40	40	40	40	40	40	40	40	40	40	40
30	68	68	68	68	68	68	68	68	68	68	46	46	46	46	46	46	46	46	46	46	46	46
31	68	68	68	68	68	68	68	68	68	68	48	48	48	48	48	48	48	48	48	48	48	48
32.1	60	60	60	60	60	60	60	60	60	60	50	50	50	50	50	50	50	50	50	50	50	50
32.2																	50					50
33																	50	50				
34													40			40		40	40	40	40	40
36	43	43	43	43	43			43		43	38	38	38	38	38	38	38	38	38	38	38	38
37														38								
39	60	60	60	60	60	60	60	60	60	60	35	35	35	35	35	35	35	35	35	35	35	35
40.1	57	57	57	57	57	57	57	57	57	57	40	40	40	40	40	40	40	40	40	40	40	40
40.2										57		40	40	40				40	40	40	40	
41																40			40		40	
44												40										

TABLE E.10. Values of NO_x after time and district.

References

Abramowitz, M. & Stegun, I. (1972). *Handbook of mathematical functions*, Dover.

Anderson, T. (1963). Asymptotic theory for principal component analysis, *The Annals of Statistics* **34**: 122–148.

Asimov, D. (1985). The grand tour: A tool for viewing multidimensional data, *SIAM Journal of Scientific and Statistical Computations* **6**(1): 128–143.

Baringhaus, L. & Henze, N. (1989). A consistent test for multivariate normality based on the empirical charateristic function, *Metrika* **35**: 339–348.

Belsley, D., Kuh, E. & Welsch, R. (1980). *Regression Diagnostics*, Whiley.

Breiman, L. & Friedman, J. (1985). Estimating optimal transformations for multiple regression and correlations, *Journal of the American Statistical Association* **80**: 580–619.

Breiman, L., Friedman, J., Olshen, R. & Stone, C. (1984). *Classification and Regression Trees*, Chapman & Hall.

Buja, A., Asimov, D. & Hurley, C. (1989). Methods for subspace interpolation in dynamic graphics, *Technical memorandum*, Bellcore.

Buja, A., Cook, D., Asimov, D. & Hurley, C. (1996). Theory and computational methods for dynamic projections in high-dimensional data visualization, *(submitted)*.

Buys, C. (1995). Spline de lissage et Spline de moindre carrés sur XploRe, (unpublished manuscript).

Chernoff, H. (1973). The use of faces to represent points in k-dimensional space graphically, *Journal of the American Statistical Association* **68**: 361–368.

Cleveland, W. (1985). *The elements of Graphing Data*, Wadsworth.

Cleveland, W., McGill, M. & McGill, R. (1986). The shape parameter of a two-variable graph, *ASA Proceedings in Statistics and Graphics*, pp. 1–10.

Cook, D., Buja, A. & Cabrera, J. (1993). Projection pursuit indexes based on orthonormal function expansions, *Journal of Computational and Graphical Statistics* **2**(3): 225–250.

Cook, D., Buja, A., Cabrera, J. & Hurley, C. (1995). Grand tour and projection pursuit, *Journal of Computational and Graphical Statistics.* (submitted).

Cook, D. & Cabrera, J. (1992). Projection pursuit indices based on fractal dimension, *Proceedings of the 24th Symposium on the Interface between Computational Science and Statistics.*

Cox, T. & Cox, M. (1994). *Multidimensional scaling*, Chapman and Hall.

Daalhuis, O. (1992). Computing with daubechie's wavelet.

Daubechie, I. (1992). *Ten Lectures on Wavelets*, CMBS-NSF, Regional Conference Series in Applied Mathematics.

Diaconis, P. & Freedman, D. (1984). Asymtotics of graphical projection pursuit, *The Annals of Statistics* **12**(3): 793–815.

Donoho, D. & Johnstone, I. (1989). Projection-based approximation and a duality with kernel methods, *The Annals of Statistics* **17**(1): 58–106.

Donoho, D., Johnstone, I., Kerkyacharian, G. & Picard, D. (1995). Wavelet shrinkage: asymptopia ? (with discussion), *Journal of the Royal Statistical Society B* **57**: 301–369.

du Toit, S., Steyn, A. & Stumpf, R. (1986). *Graphical Exploratory Data Analysis*, Springer Verlag.

Duan, N. & Li, K. (1991). Slicing regression: A link-free regression method, *The Annals of Statistics* **19**(2): 505–530.

Earp, L. & Rotermund, H. (1987). Prüfstand für Mikroprozessoren, *Chip* pp. 154–156.

Fan, J., Gasser, T., Gijbels, I., Brockmann, M. & Engel, J. (1993). Local polynomial fitting: A standard for non-parametric regression, *Discussion paper 9315*, Institute of Statistics, Catholic university of Louvain.

Fan, J. & Gijbels, I. (1996). *Local Polynomial Modeling and Its Application—Theory and Methodologies*, Chapman & Hall.

Fan, J. & Marron, J. (1994). Fast implementation of nonparametric curve estimators, *Journal of Computational and Graphical Statistics.*

Fisherkeller, M., Friedman, J. & Tukey, J. (1988). Prim-9 : An interactive multidimensional data display and analysis system, *Collected Works of John W. Tukey, Volume V, Graphics: 1965-1985*, Pacific Grove, pp. 308–327.

Flury, B. & Riedwyl, H. (1981). Graphical representation of multivariate data by means of asymmetrical faces, *Journal of the American Statistical Association* **76**(376): 757–765.

Friedman, J. (1985). Classification and multiple regression through projection pursuit, *Technical report LCS012*, Department of Statistics, Stanford University.

Friedman, J. (1987). Exploratory projection pursuit, *Journal of the American Statistical Association* **82**: 249–266.

Friedman, J. & Stuetzle, W. (1981a). Projection pursuit classification, (unpublished manuscript).

Friedman, J. & Stuetzle, W. (1981b). Projection pursuit regression, *Journal of the American Statistical Association* **76**: 817–823.

Friedman, J., Stuetzle, W. & Schroeder, A. (1984). Projection pursuit density estimation, *Journal of the American Statistical Association* **79**: 599–608.

Friedman, J. & Tukey, J. (1974). A projection pursuit algorithm for exploratory data analysis, *IEEE Transactions on Computers C* **23**: 881–889.

Gellert, W., Küstner, H., Hellwich, M. & Kästner, H. (1977). *Kleine Enzyklopädie Mathematik*, VEB Bibliographisches Institut Leipzig.

Germer, T. & Neudeck, R. (1994). *Betrachtungen des Angebotes für Wohnungseigentum in Berlin*, Master's thesis, Institut für Statistik und Ökonometrie, Wirtschaftwissenschaftliche Fakultät der Humboldt-Universität zu Berlin.

Hall, P. (1989a). On polynomial-based projection indices for exploratory projection pursuit, *The Annals of Statistics* **17**(2): 589–605.

Hall, P. (1989b). On projection pursuit regression, *The Annals of Statistics* **17**(2): 573–588.

Härdle, W. (1990). *Applied nonparametric regression*, Cambridge University Press.

Härdle, W. (1991). *Smoothing techniques: With implementation in S*, Springer Verlag.

Härdle, W., Kerkyachrian, G., Picard, D. & Tsybakov, A. (1995). Wavelets and econometric applications, (unpublished manuscript).

Härdle, W. & Simar, L. (1995). Applied multivariate analysis, (unpublished manuscript).

Harrison, D. & Rubinfeld, D. (1978). Hedonic prices and the demand for clean air, *Journal for Environmental Economics & Managment* **5**: 81–102.

Hartigan, J. (1975). *Clustering techniques*, New York: Whiley.

Huber, P. (1985). Projection pursuit, *The Annals of Statistics* **13**: 435–475.

Inselberg, A. (1985). The plane with parallel coordinates, *The Visual Computer* **1**: 69–91.

Jambu, M. & Lebeaux, M. (1983). *Cluster Analysis and Data Analysis*, North-Holland.

Jee, J. (1985). Exploratory projection pursuit using nonparametric density estimation, *ASA. Proc. of Stat'l. Computing Sect.*, pp. 335–339.

Jones, M. & Sibson, R. (1987). What is projection pursuit ?, *Journal of the Royal Statistical Society A* **150**: 1–18.

Klinke, S. (1994). A fast implementation of kernel-based projection pursuit indices, *Discussion paper 9404*, Institute of statistics, Catholic university of Louvain.

Klinke, S. & Cook, D. (1995). Implementation of kernel based indices in XGobi, *Discussion paper 47*, SFB 373, Humboldt-University of Berlin.

Koch, A. & Haag, U. (1995). The statistical software guide '94/95, *Computational Statistics & Data Analysis* **19**(2): 237–261.

Kolev, N. (1993). *Applied Statistics 1*, Stopanstvo, Sofia.

Kötter, T. & Turlach, B. (1995). Additive modeling, *in* W. Härdle, S. Klinke & B. Turlach (eds), *XploRe - an interactive statistical computing environment*, Springer, pp. 223–249.

Krishnan, R., Müller, R. & Schmidt, P. (1995). Accessing "computable" information over WWW: the MMM project, *in* T. X. Bui (ed.), *Proceedings of the Third International Society for Decision Support Systems Conference*, International Society for DSS, Hong Kong.

Kruskal, J. (1964a). Multidimensional scaling by optimzing goodness-of-fit to nonmetric hypothesis, *Psychometrika* **29**: 1–27.

Kruskal, J. (1964b). Nonmetric multidimensional scaling: a numerical method, *Psychometrika* **29**: 115–129.

Kruskal, J. (1969). Toward a practical method which helps uncover structure of a set of multivariate observations by finding the linear transformation which optimizes a new "index of condensation", *Statistical Computation*, Academic Press, pp. 427–440.

Kruskal, J. (1972). Linear transformation of multivariate data to reveal clustering, *Multidimensional Scaling : Theory and Applications in the Behavioural Scienes*, Seminar Press.

Li, K. (1991). Sliced inverse regression for dimension reduction (with discussion), *Journal of the American Statistical Association* **86**(414): 316–342.

Li, K. (1992). On principal hessian directions for data visualization and dimension reduction: Another application of stein's lemma, *Journal of the American Statistical Association* **87**(420): 1025–1039.

Li, K., Aragon, Y. & Thomas-Agnan, C. (1995). Analysis of multivariate outcome data: Sir and a nonlinear theory of hotelling's most predictable variates, *Discussionpaper*, Université de Toulouse, GREMAQ.

Lütkepohl, H. (1990). Skript zur Vorlesung multiple Zeitreihenanalyse.

Mardia, K. (1980). *Tests of univariate and multivariate normality*, Vol. 1 of *Handbook of statistics*, North-Holland Publishing Company, pp. 279–320.

Mardia, K., Kent, J. & Bibby, J. (1979). *Multivariate Analysis*, Academic Press.

McGill, R., Tukey, J. & Larsen, W. (1978). Variations of box plots, *The American Statistician* **32**: 12–16.

Meyer, C. (1994). Kontrastprogramm - Sind Power Macs die besseren PCs ?, *c't* pp. 140–143.

Meyer, C. & Siering, P. (1994). Ausgezählt - Power Macs gegen Intel PCs: Zahlen und Fakten, *c't* pp. 143–145.

Morrison, D. (1976). *Multivariate Statistical Methods*, McGraw-Hill.

Morton, S. (1989). *Interpretable Projection Pursuit*, PhD thesis, Department of Statistics, Stanford University.

Mucha, H. (1992a). *Clusteranalysis auf Mikrocomputern*, Akademie Verlag.

Mucha, H. (1992b). Specific metrics for cluster analysis and principal components analysis, *SoftStat '91: Advances in Statistical Software*, Gustav Fisher, pp. 249–258.

Mucha, H. & Klinke, S. (1993). Clustering techniques in the interactive statistical computing environment `xplore`, *Discussion paper 9318*, Institute of Statistics, Catholic university of Louvain.

Nachtmann, L. (1987a). Der 32-bit-Mikroprozessor, *Chip* **2**(2): 48–49.

Nachtmann, L. (1987b). Der grosse Benchmark, *Chip* **2**(10): 318–319.

Neumann, M. (1994). Spectral density estimation via nonlinear wavelet methods for stationary non-gaussian time series, *Statistics research report no. SRR 028-94*, CMA, Australian National University, Canberra.

Oliver, G., Müller, R. & Weigend, A. (1995). The design of MMM: A Model ManageMent system for time series analysis, *in* J. Ford, F. Makedon & S. A. Rebelsky (eds), *Electronic Publishing and the Information Superhighway (DAGS95)*, Birkhäuser, Boston, Basel, Berlin, pp. 223–233. On-line conference proceedings: Addison-Wesley Interactive http://www.aw.com/awi.html.

Pearson, K. (1901). On lines and planes of closest fit to systems of points in space, *Philosophical Magazine* **6**: 559–572.

Polzehl, J. (1995). Projection pursuit discriminant analysis, *Computational Statistics & Data Analysis.*

Polzehl, J. & Klinke, S. (1994). Experiences with bivariate projection pursuit indices, *Discussion paper 9416*, Institute of Statistics, Catholic university of Louvain.

Polzehl, J. & Klinke, S. (1995). Exploratory projection pursuit, *in* W. Härdle, S. Klinke & B. Turlach (eds), *XploRe - an interactive statistical computing environment*, Springer, pp. 169–194.

Posse, C. (1995a). Projection pursuit exploratory data analysis, *Journal of the American Statistical Association* p. (submitted).

Posse, C. (1995b). Which tools for exploratory projection pursuit ?, *Journal of Computational and Graphical Statistics* p. (submitted).

Press, W., Flannery, B., Teukolsky, S. & Vetterling, W. (1988). *Numerical recipes in C*, Cambridge University Press.

Proenca, I. (1994). A teachware module to instruct statistics and econometrics in XploRe 3.1, *Discussion paper 9408*, Institute of Statistics, Catholic university of Louvain.

Proenca, I. (1995). Interactive graphics for teaching simple statistics, *XploRe - an interactive statistical computing environment*, Springer, pp. 113–140.

Rand, W. (1971). Objective criteria for the evaluation of clustering methods, *Journal of the American Statistical Association* **66**: 846–850.

Rousseeuw, P. (1994). Least median of squares regression, *Journal of the American Statistical Association* **79**: 871–880.

Schmidt, G., Mattern, R. & F., S. (1981). Biomechanical investigation to determine physical and traumatological differentiation criteria for the maximum load capacity of head and vertebral column with and without protective helmet under effects of impact, *EEC research program on biomechanics of impacts.*, Final Report Phase III, Project 65, Institut für Rechtsmedizin, Universität Heidelberg.

Schneider, B. (1994). Querleser - Zerstreutheit mit System: Skip-Listen,, *c't* **2**: 204–207.

Schnurer, G. (1992). Zahlenknacker im Vormarsch, *c't* **10**(4): 170–186.

Schwab, G. (1991). *Fehlende Werte in der angewandten Statistik*, Deutscher Universitätsverlag.

Scott, D. (1985). Averaged shifted histograms: Effective nonparametric density estimation in several dimensions, *The Annals of Statistics* **13**: 1024–1040.

Scott, D. (1992). *Multivariate density estimation*, John Wiley & Sons.

Seifert, B., Brockmann, M., Engel, J. & Gasser, T. (1994). Fast algorithms for nonparametric curve estimation, *Journal of Computational and Graphical Statistics* **3**(2): 192–213.

Silverman, B. (1984). A fast and efficient cross-validation method for smoothing parameter choice in spline regression, *Journal of the American Statistical Association* **79**: 584–589.

Silverman, B. (1986). *Density estimation*, Chapman and Hall.

Sneath, P. (1957). The application of computers to taxonomy, *Journal of Genetics and Microbiology* **17**: 201–226.

Sokal, R. & Michener, C. (1958). A statistical method for evaluating systematic relationships, *University of Kansas Science Bulletin* **38**: 1409–1438.

Stuetzle, W. (1987). Plot windows, *Journal of the American Statistical Association* **82**(398): 466–475.

Theilen, B. (1990). *Tests auf multivariate Normalverteilung*, Master's thesis, Institut für Statistik und Ökonometrie der Universität Kiel.

Theus, M. (1996). Software review: Data desk 5.0, *Computational Statistics* **11**(2): 205–210.

Tukey, J. (1970). *Exploratory Data Analysis – Limited Preliminary Edition*, Addison-Wesley.

Tukey, J. (1990). Data-based graphics: Visual display in the decades to come, *Statistical Science* **5**(3): 327–339.

Unwin, A. (1995). Manet - missings are now equally treated, Talk at the Workshop "Data Analysis", Heidelberg, Germany.

Ward, J. (1963). Hierarchical grouping to optimize an objective function, *Journal of the American Statistical Association* **58**: 236–244.

Wegman, E. (1990). Hyperdimensional data analysis using parallel coordinates, *Journal of the American Statistical Association* **85**(411): 664–675.

Wilson, S. (1982a). Sound and exploratory analysis of contingency table data, *2 COMPSTAT, 2nd. Symp. on Computational Statistics* **83**: 243–250.

Wilson, S. (1982b). Sound and exploratory data analysis, *2 COMPSTAT, 2nd. Symp. on Computational Statistics* **5**: 447–450.

Wolfram, S. (1991). *Mathematica - A System for Doing Mathematics by Computer*, Addison-Wesley.

Yenjukov, I. (1989). *Indeces for Projection Pursuit*, Nova Science Publisher, New York, pp. 181–188.

GPSR Compliance

The European Union's (EU) General Product Safety Regulation (GPSR) is a set of rules that requires consumer products to be safe and our obligations to ensure this.

If you have any concerns about our products, you can contact us on ProductSafety@springernature.com

In case Publisher is established outside the EU, the EU authorized representative is:

Springer Nature Customer Service Center GmbH
Europaplatz 3
69115 Heidelberg, Germany

Batch number: 10577878

Printed by Printforce, the Netherlands